Sex Steroids and Apoptosis In Skeletal Muscle: Molecular Mechanisms

Edited by

Andrea Vasconsuelo

Instituto de Ciencias Biológicas y Biomédicas del Sur (INBIOSUR), Universidad Nacional del Sur- CONICET, Bahía Blanca, Argentina

Sex Steroids and Apoptosis In Skeletal Muscle: Molecular Mechanisms

Editor: Andrea Vasconsuelo

ISBN (Online): 978-9-81141-236-3

ISBN (Print): 978-9-81141-235-6

© 2019, Bentham eBooks imprint.

Published by Bentham Science Publishers Pte. Ltd. Singapore. All Rights Reserved.

need for a court order if at any point you breach any terms of this License Agreement. In no event will any delay or failure by Bentham Science Publishers in enforcing your compliance with this License Agreement constitute a waiver of any of its rights.

3. You acknowledge that you have read this License Agreement, and agree to be bound by its terms and conditions. To the extent that any other terms and conditions presented on any website of Bentham Science Publishers conflict with, or are inconsistent with, the terms and conditions set out in this License Agreement, you acknowledge that the terms and conditions set out in this License Agreement shall prevail.

Bentham Science Publishers Pte. Ltd.
80 Robinson Road #02-00
Singapore 068898
Singapore
Email: subscriptions@benthamscience.net

BENTHAM SCIENCE

CONTENTS

FOREWORD

Aging has serious consequences on skeletal muscle. 'Sarcopenia', the progressive loss of muscle mass and associated muscle weakness during elderly, affects radically the functional capacity and general health of adult people and renders frail elders susceptible to serious injury from sudden falls and fractures and at risk for losing their functional independence. There is a vital need to recognise the molecular mechanisms and regulatory factors, underlying age-related muscle wasting and to develop therapeutic strategies that can attenuate, prevent, or finally reverse sarcopenia. In this context, sexual hormones play a key role.

The book, *SEX STEROIDS AND APOPTOSIS IN SKELETAL MUSCLE: MOLECULAR MECHANISMS,* written by Dr. Andrea Vasconsuelo aims to provide a new way to perceive the role of sex hormones in skeletal muscle. The book presents, in integrated form, the latest information on sarcopenia and its relation with apoptosis, by leading researchers, studying the cellular and molecular mechanisms underlying age-linked changes in the skeletal muscles emphasising on the role of satellite cells. The authors succeed to explain, how the hormones are involved in muscle homeostasis and in the regulation of apoptosis process, and how these two muscles functions connect to maintain a healthy muscle or to trigger pathologies, Therefore, the goal of this work is to highlight the combination of that information focusing on the molecular level and resulting in to highlight the clarification of molecular mechanisms implicated in skeletal muscle aging, showing that when apoptosis is more intense, sex hormones levels decline. Very interesting, the book, showing that contains a chapter describing molecular structure of phytoestrogens and their action on sex steroids receptors. In addition, it is also important to emphasize the drafting and writing of the book promoting easy and pleasant reading, with careful documentation and high quality images of the ebook experiments and comprehensive and updated bibliography. This ebook is of interest to graduates and postgraduates in the fields of medicine and biochemistry, to researchers of different fields of ageing biology and people of the pharmaceutical, to health-care industry fields.

Dr. Anabela La Colla
Dpto. De Química,
Facultad de Ciencias Exactas y Naturales,
Universidad Nacional De,
Mar del Plata. Argentina

PREFACE

There are few ebooks available contains details on the integration of cellular apoptosis and sex steroids at the molecular level. Here, our intention is to show how hormonal signals activate cellular responses that have been discovered individually but, in reality when a signal reaches a cell those signaling cascades interact with all cellular components to different degrees. Therefore, with this work, we want to present in an integrated way the concepts of hormonal regulation and apoptosis. Moreover, it is important to know how these processes are at the molecular level since mistakes in them lead to pathologies. The present work centers on the role of 17b-Estradiol (E2) and Testosterone (T) in most animal tissues, in addition to the reproductive system. Highlighting the role of both hormones in the lifespan of the skeletal muscle cell.

First, we describe the role of E2 and T affecting growth and cell functions in mammals. Accordingly, the nuclear estrogen (ER) and androgen (AR) receptors are ubiquitously expressed. Moreover, ER and AR may have non-classical intracellular localizations, e.g. plasma membrane, mitochondria, and endoplasmic reticulum, raising complexity to the actions of E2 and T. As well as genomic actions, sex steroids can fast regulate signaling pathways by non-genomic mechanisms through ER and AR, too.

We continue describing basic concepts of programmed cell death and how both sex hormones can regulate apoptosis through those signaling pathways. In mitochondria, the existence of ER and AR and actions of estrogen and androgen have been demonstrated, in keeping with the organelle being the main switch point of programmed cell death. The recurrent action for each steroid hormone is the safeguard of mitochondria against diverse insults, resulting in cellular survival. Then we explain the role of sex hormones in mitochondrial physiology (ROS production, regulation of mitochondrial enzymes and oxidative system pathway). In addition, we describe the action of sex hormones on muscle stem cells at the molecular level. The book shows how the integration of all the processes described (the effects of sex hormones, mitochondrial dysfunction, increased apoptosis, depletion in muscle stem cells, augmented production of cellular toxins as ROS, and anomalous regulation of stress pathways) results in sarcopenia. Sarcopenia is a predominant disorder among the elderly, which implies the loss of muscle mass and strength. Although the basis of sarcopenia is unclear, evidence suggests that the putative molecular mechanism associated with this condition could be apoptosis. Remarkably, sarcopenia has been linked to a deficit of sex hormones, which decrease upon aging. The skeletal muscle capability to repair and regenerate itself would not be possible without satellite cells, a subpopulation of cells that remain quiescent throughout life. In response to stress, this muscle stem cells are activated directing skeletal muscle regeneration. Of relevance, satellite cells are E2 and T target. To end, the last chapter is devoted to natural compounds that have similarity with sex hormones and that could, therefore, have therapeutic potential. Finally, I thank the people who collaborated in this work and I hope that it fulfills the objective of presenting an integrated vision of the cellular mechanisms that are activated in response to hormones, regulating apoptosis specifically in the skeletal muscle.

Andrea Vasconsuelo
Instituto de Ciencias Biológicas y Biomédicas del Sur (INBIOSUR),
Universidad Nacional del Sur- CONICET,
Bahía Blanca,
Argentina

List of Contributors

Andrea Vasconsuelo	Instituto de Ciencias Biológicas y Biomédicas del Sur (INBIOSUR), Universidad Nacional del Sur- CONICET, Bahía Blanca, Argentina
Florencia Antonella Musso	Instituto de Química del Sur (INQUISUR, Universidad Nacional del Sur-CONICET, Bahía Blanca, Argentina
Lucía Pronsato	Instituto de Investigaciones Biológicas y Biomédicas del Sur (INBIOSUR CONICET-UNS), Bahía Blanca, Argentina
Lorena M. Milanesi	Instituto de Ciencias Biológicas y Biomédicas de Sur (INBIOSUR)-UNS-CONICET, Bahía Blanca, Argentina
Lucía Pronsato	Instituto de Ciencias Biológicas y Biomédicas del Sur (INBIOSUR), Universidad Nacional del Sur- CONICET, Bahía Blanca, Argentina
María Belén Faraoni	Instituto de Química del Sur (INQUISUR, Universidad Nacional del Sur-CONICET, Bahía Blanca, Argentina

CHAPTER 1

Subcellular Localization and Physiological Roles of Androgen Receptor

Lucía Pronsato[*]

Instituto de Investigaciones Biológicas y Biomédicas del Sur (INBIOSUR CONICET-UNS), 8000 Bahía Blanca, Argentina

Abstract: Androgens, such as testosterone and Dihydrotestosterone (DHT), exert their actions through the Androgen Receptor (AR), a ligand-dependent nuclear transcription factor that belongs to the steroid hormone nuclear receptor superfamily. The actions of androgens can be mediated through the AR in a DNA binding-dependent manner to modulate the transcription of target genes, or in a manner independent of DNA binding, to trigger rapid cellular events such as the activation of the second messenger signaling pathway. The AR is expressed ubiquitously and it has a wide variety of biological actions comprising significant roles in the development and maintenance of the reproductive, skeletal muscle, cardiovascular, immune, neural and haemopoietic systems, exerting a diversity of roles in many physiological and pathological processes. Studies with AR Knockout (ARKO) mouse models, specifically the cell type- or tissue-specific ARKO models, have revealed many cell type- or tissue-specific pathophysiological roles of AR in mice. Because of the huge amount of information about androgens and the AR, this chapter is not presented as an extensive review of all of it, but rather as an overview of the expression and biological function of AR as well as its significant role in clinical medicine.

Keywords: Androgens, Androgen Receptor, Biological Action, Tissue Distribution.

INTRODUCTION

Adequate regulation of androgens action is essential for a variety of developmental and physiological processes, mainly male sexual development and maturation, male reproductive organs maintenance and spermatogenesis [1 - 4]. Similarly, androgens are central in the functioning of several other organs and tissues. The major physiological androgens, Testosterone (T) and its metabolite

[*] **Corresponding author Lucía Pronsato:** Instituto de Investigaciones Biológicas y Biomédicas del Sur (INBIOSUR CONICET-UNS) 8000, Bahía Blanca, Argentina; Tel: +54 291 4595101x4337; E-mail: lpronsato@criba.edu.ar

Andrea Vasconsuelo (Ed.)

5α-dihydrotestosterone (DHT), mainly mediate their biological actions through binding to the androgen receptor (AR).

AR is a member of the nuclear receptors (NR) superfamily, a group of transcription factors that trigger the transcription of their target genes in response to specific ligands [5, 6]. They are implicated in a biological process such as development, differentiation, reproduction and homeostasis of eukaryotic organisms. NRs have been preserved during evolution [7], and can be divided into three classes: type I receptors are steroid receptors that include the AR, estrogen receptor (ER), progesterone receptor (PR), mineralocorticoids receptor (MR) and glucocorticoids receptor (GR), classically defined as ligand-dependent, that homodimerize to exert their function. The type II nuclear receptors are known as the retinoid-thyroid family, and consist of the receptors for vitamin D (VDR), thyroid hormone (TR), retinoic acid (RAR), and the peroxisome proliferator-activated receptors (PPAR); they are ligand-independent with potential to both homodimerize and heterodimerize [8]. Finally, the receptors of the third class, named Orfan, comprise a group of proteins that share sequences with significant homologies, whose ligands have not been characterized [9, 10]. The comparative functional and structural analyzes of the NRs revealed that they contain a similar structural organization and can be divided into four functional domains: the carboxyl-terminal ligand binding domain (LBD) is connected by a hinge region (H) to a highly conserved DNA-binding domain (DBD). The LBD includes a hormone-dependent coactivator interface named activation function 2 (AF2). The amino-terminal domain (NTD) contains a hormone-independent coactivator interface, AF1. It is the least conserved domain and has little intrinsic structure. The binding to the DNA or the interaction with other proteins leads to a more ordered structure [11, 12]. The NTD encloses the majority of phosphorylation sites, many of which have serine-proline motifs (Ser-Pro) which can be recognized by the peptidyl-prolyl isomerase, Pin1 [13]. Therefore, phosphorylation of these sites can lead to the isomerization and thus, alteration of the structure of the receptor.

Structural Organization of the AR Gene

The gene that encodes for AR is found on the long arm of chromosome X (Xq11.2-12), and was discovered in 1981, when it was genetically studying humans and mice that showed androgen insensitivity [14 - 16]. In 1988, the AR cDNA was cloned for the first time, in spite of the difficulties to obtain enough quantities of the purified protein, to produce antibodies or partial amino acid sequences to design synthetic oligonucleotide probes [5, 17]. The AR gene size is around 90 Kb and contains 8 exons, and its structural organization is almost identical to the genes that encode for the other members of steroid hormone receptors, suggesting a common ancestral past [18, 19]. The possibility of the existence of additional AR genes, which encode for an AR with unclassical localization in the plasma membrane, has been suggested. This idea firstly arose from the observation of effects triggered by testosterone, at short times of hormonal treatment (responses within few minutes or seconds) that could not be a consequence of the transcriptional activity of the classical AR, in brain and osteoblasts [20, 21]. Although only one gene for AR has been detected in humans, two isoforms of AR mRNA were found in the male larynx of Xenopus laevis [22]. Since a second gene encoding for the estrogen receptor has been found, it is possible that other members of the steroid receptor superfamily have also multiple isoforms of the encoding gene.

Protein Structure of AR

The AR protein consists of approximately 919 amino acids and a molecular weight of 98 kDa, which are structured in 4 functional domains [23]. The N-terminal regulatory domain, encoded mainly by exon 1 (1-555 bp), mediates transcriptional activity. This domain contains the activation region of the ligand-independent transcription (AF-1), being this place a site of interaction with certain co-regulators. The DBD, encoded by exons 2 and 3 (556-623 bp), contains two zinc fingers capable of interacting specifically with small sequences named androgen response elements (AREs). The hinge region, encoded by exon 4 between 624-665 bp, is important for the receptor movement. Finally, the LBD encoded by the last exons 5, 6, 7 and 8 between 666-918 bp, is the place where the androgens bind to the receptor, and contains the activation region of the ligand-dependent transcription (AF-2) [24, 25]. The DBD and the LBD share a high grade of homology with the other steroid receptors.

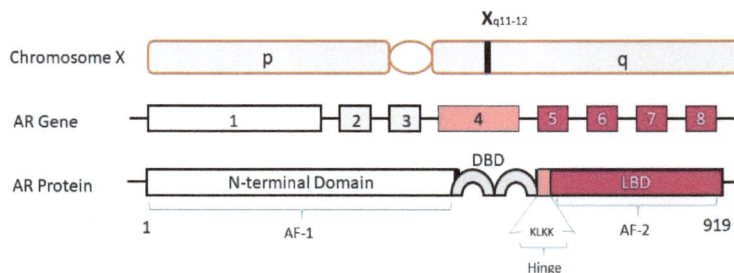

Schematic representation of the human AR gene and protein. The AR gene is localized on the long arm of the X chromosome. It is encoded by 8 exons (919 amino acids) and the protein contains different structural domains: the N-terminal domain (NTD) that includes the activation function 1 (AF1) domain, the DNA binding domain (DBD), the ligand binding domain (LBD) that contains the activation function 2 (AF2) domain and the hinge region containing the KLKK motif.

Classical Mechanism of Action of AR

In the absence of ligand, AR is found in a monomeric form, making a complex with heat shock proteins (Hsp), such as Hsp90, Hsp70 and Hsp56 [26] that act as chaperones. An essential function of the Hsp heterocomplex is to enable folding of the LBD into a high-affinity steroid-binding conformation. Hsp90 modulates hormone binding affinity *in vivo* [27], and Hsp90s are necessary for the acquirement of active conformation in agonist-bound AR to modulate nuclear transfer, nuclear matrix binding, and transcriptional activity [28]. This complex is dynamic and can translocate between the cytoplasm and the cell nucleus, although the relative subcellular distribution in absence of ligand is mainly cytoplasmic. LBD contains ligand-dependent activation function AF-2. Agonist binding provokes a conformational modification particularly in the C-terminal AF-2, which exposes an amphipathic α-helix to interact with coactivator proteins. Several coactivators bind to a surface formed by helices 3, 4, and 12, and the relocation of helix 12 is central for this interaction. AR is unique among steroid receptors, since its N-terminal AF-1 is crucial in the transcriptional activation, and an LBD-deficient AR is constitutively active [29, 30].

Given that androgens are lipid hormones derived from cholesterol, they are able to diffuse freely through the plasma membrane. The hormone (T or DHT) binding to the AR LBD produces a series of conformational changes in the receptor (activated state) that causes the release of the heat shock proteins and allows a coactivator binding pocket presentation, letting the docking of coactivator proteins [31, 32].

The binding of the hormone, provokes modifications that result in the receptors transportation from the cytosol to the nucleus [33, 34]. In the nucleus, receptors are further compartmentalized to subnuclear domains associating and dissociating with chromatin and the nuclear matrix [35, 36]. Steroid receptors (SRs) shuttle between the cytosol and the nucleus and the balance between export and import defines their location [37]. Nuclear localization signal (NLS) is necessary for nuclear targeting of proteins. After hormone removal, the export of steroid receptors from the nucleus usually occurs in several hours, despite the release of hormone from receptor takes place very quickly [38].

Members of the NRs superfamily directly activate or repress target genes by binding to hormone response elements (HREs) in the promoter or enhancer areas of the genes [39, 40]. Steroid receptors bind as ligand-induced receptor dimers to HREs consisting in numerous cases of two inverted 6bp half-sites separated by three nucleotides. HREs confer specificity to receptor dimer binding [41], and the spacer nucleotides and the areas flanking the half sites, exert a central role in the establishment of receptor binding specificity [42]. When the receptor is not bound to the ligand, it is weakly associated with DNA. Once the ligand binding, the AR interacts in a stable way with DNA sequences named androgen response elements (AREs), which are characterized by a repeated consensus sequence 5'-TGTTCT-'3 that localizes in the promoter and enhancer of genes that respond to androgens region [16]. This association provoked the recruitment of other proteins named coregulators (coactivators or corepressors), that are required to activate or repress the transcription of target genes [40]. The interaction of AR with specific AREs is necessary for androgen-dependent transcriptional activation, while DNA binding is not generally needed for transrepression by AR [43, 44]. Next level of steroid receptor selectivity is provided by the highly variable N-terminal areas of the steroid receptors that are responsible for steroid-specific modulation of specific genes [45]. Also areas near the AR DBD is crucial for the DNA-binding specificity [46].

Non-Classical Mechanisms of Action of AR

Additionally to the classical genomic mechanism of action of steroids, which involves the activation of intracellular NRs, there is wide evidence that steroids also activate receptors on the cell surface to induce quick intracellular signaling and biological responses, that are frequently non-genomic.

As described above, the steroid hormones bind to receptors present in the nucleus or cytoplasm, and then, the receptor-ligand complex translocates to the nucleus where it modulates the gene transcription and protein synthesis. This is a slow process that requires at least 30 or 40 minutes to modify the genes expression at the transcriptional level, and several hours to produce significant changes in the levels of the newly synthesized proteins [47]. However, non-genomic actions exerted by hormones through their receptors, have been described [48 - 50]. These actions occur in a few seconds or minutes, after the addition of the agonist.

Numerous studies suggest that androgens modulate cellular processes through a non-classical mechanism [51 - 59]. These non-classical actions involve the rapid stimulation of signaling cascades inducing the increase of intracellular Ca^{2+} levels, cAMP, activation of MAPK, PKA, PKC and the 3-phosphoinositol protein kinase pathway (PI3K/Akt). Different from the classical genomic mechanism of action of androgens, these events are not suppressed by the inhibition of transcription and can be induced in some cells or tissue types employing macromolecular derivatives of the hormone, that are not permeable to the plasma membrane, which would point to the existence of membrane entities, which are able to bind the hormone [60]. These quick effects are thought to be non-genomic, given that they take place in cells that do not have functional ARs or they are considered to

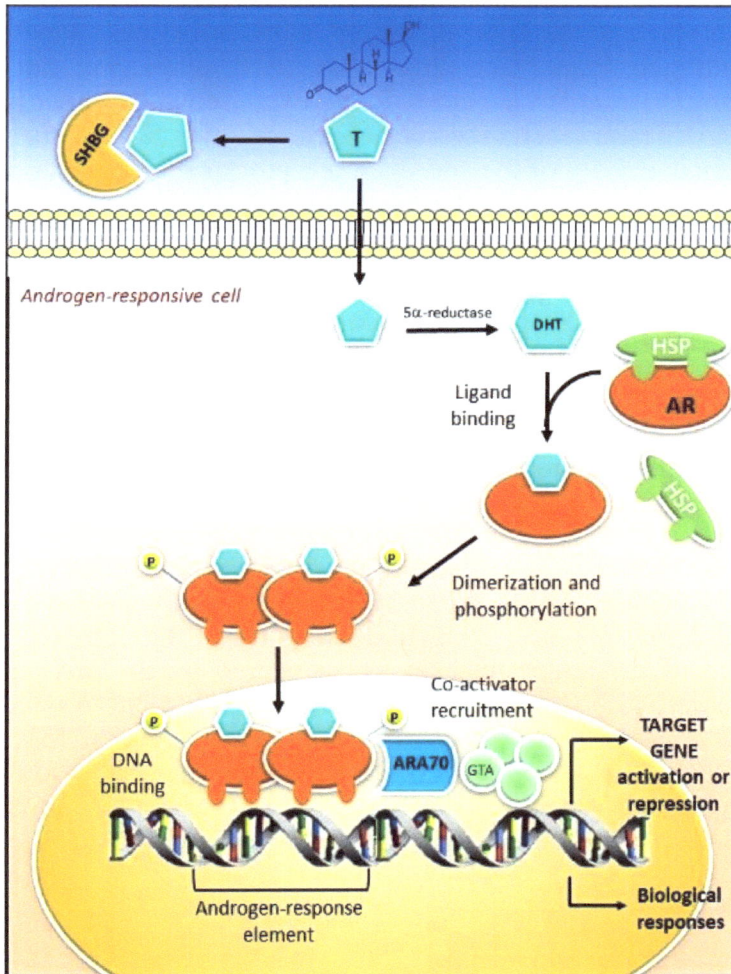

Classical mechanism of action of androgens. Testosterone (T) circulates in the blood associated to sex-hormone-binding globulin (SHBG) as well as to albumin, and exchanges with free testosterone. Free T enters androgen-responsive cells and can be converted to dihydrotestosterone (DHT) by the 5α-reductase enzyme or exerts its action as T. The T/DHT binding to the androgen receptor (AR) leads to dissociation from heat-shock proteins (HSPs) and the phosphorylation of the AR. The receptor forms dimers and can bind to androgen-response elements (ARE) in the promoter areas of target genes. Co-activators (ARA70) and co-repressors (not shown) also bind the AR complex, allowing or avoiding, respectively, its association with the general transcription apparatus (GTA). Activation or repression of specific genes conduces to biological responses including growth, developing, survival and proliferation.

be mediated through an AR functioning on cell surface or in the cytosol to activate the mitogen-activated protein kinase (MAPK) signal cascade [61]. It has been suggested that cytoplasmic AR may interact with the protein tyrosine kinase c-Src and cause the rapid activation of MAPKs and the PI3K/Akt pathway, regulating thus, the intracellular signaling cascades of these kinases [51, 62]. Moreover, androgens could act through the SHBG receptor and possibly a different G protein-coupled receptor to activate second messenger signaling mechanisms [63]. These second messenger cascades may finally attend to regulate the transcriptional activity of the androgen receptor or other transcription factors. AR, PR and the ER are capable of activating the MAPK through a non-genomic mechanism independent of their transcriptional activity [51, 64, 65].

Additionally to the classic androgen receptor, androgens can also induce second messenger cascades through at least one plasma membrane receptor. Membrane receptor-mediated effects are characteristically not suppressed by antagonists of the classical androgen receptor, and they can be detected in cells devoid of AR [66, 67]. The scientific evidence accumulated, talks about the presence of membrane androgen receptors (mAR), activating fast non-transcriptional signals. Although the precise molecular identity of mAR is still not known, evidence has suggested the existence of androgen-binding sites in the plasma membrane of various cell types and tissues, like T lymphocytes, prostate cells, skeletal muscle, Sertoli cells and oocytes [54, 67 - 70]. Studies in the last years have involved key pro-survival and pro-apoptotic genes like Akt, NF-kB, Bad, Fas and caspase-3 in the modulation of the apoptotic response triggered by the mAR activation in prostate cancer cells [71]. AR, PR, and ER have been found to interact with the intracellular tyrosine kinase c-Src, inducing c-Src activation [51, 64, 65]. In LNCaP cells, inhibition of c-Src kinase or MAPK activity avoids androgen-induced cell cycle progression [51]. In spite of this emergent information, the molecular mechanisms that mediate these non-genomic actions are still poorly understood.

TISSUE DISTRIBUTION OF AR

Muscle

A positive reaction for AR has been detected in almost every nucleus of skeletal muscle. In cardiac muscle, greatest amount of nuclei of the ventricular and atrial myocardial cells express the receptor. Nuclear staining has been slightly weaker in female than in male tissues [72]. Although the presence of AR in skeletal muscles was once questioned [73, 74], skeletal muscle and levator ani muscle in the rat have AR with binding characteristics similar to the receptors in other androgen targets [75 - 79]. The higher level of cytosolic ARs in the levator ani muscle compared to other skeletal muscle in the rat could explain the greater sensitivity of this muscle to alterations in serum androgen concentrations [77].

Scientific evidence collected in the last years points to the presence of non-classical membrane androgen receptors (mAR), activating quick, non-genomic signals. While the precise molecular identity of mAR still remains unidentified, non-genomic androgen actions exhibited within minutes have been showed in numerous cell types including skeletal muscle [80]. Furthermore, extranuclear organelles have been proposed as containing sex steroid receptors. The existence of ER and AR in mitochondria of mammalian cells including skeletal muscle has been reported [80 - 86]. Thus, non-classical localization of AR from where it can exert non-genomic actions could be possible.

C2C12 murine skeletal muscle cells are myoblasts derived from satellite cells, which behave as parent lineage, and are subclones of the C2 myoblasts [87]. C2C12 myoblasts are similar to the activated satellite cells that surround the mature myofibers and proliferate and differentiate, contributing to the reparation of the tissue when a cellular injury exists [88]. It has been proved that androgens protects skeletal muscle C2C12 cells against apoptosis through a mechanism involving intermediates of the apoptotic intrinsic pathway and the androgen receptor [89]. Biochemical and immunological data has supported mitochondrial and microsomal localization of the androgen receptor in the C2C12 skeletal muscle cells. Western blot assays of subcellular fractions made possible the immunodetection of a band of a 110 kDa, probably corresponding to the classical AR, in total muscle cell homogenates and fractions derived therefrom, including mitochondria and microsomes. The androgen receptor is mainly a nuclear receptor. Nevertheless, accumulated evidence points to the existence of extranuclear AR entities, structurally and functionally similar to the well-known AR [90]. Remarkably, extra immunoreactive bands have been detected in all the subcellular fractions evaluated. The immunoreactive proteins obtained could be the result of alternative usage of different inframe initiations codons or splice variants of the full-length AR transcript, as described in other research [86, 91]. Likewise, these studies suggest the existence of functional androgen binding sites in classical and non-classical compartments. The identification and characterization of these androgen binding entities in total homogenate and subcellular fractions of the C2C12 cells were done by competitive radioligand binding assays with [^{3}H] T, which revealed specific binding activity in all the subcellular fractions, principally localized in the nuclear pool. Furthermore, the specific binding in mitochondrial and microsomal fractions isolated from C2C12 cells was saturable process with respect to the ligand concentration. Scatchard linearization of the saturation binding data was consistent with a single set of affinity binding sites. Therefore, the AR specific binding sites, detected in total homogenates, are present not only in cytoplasm and nucleus but also in other particulate subfractions [90]. This is in accordance to studies where an considerable content of ARs was detected in mitochondrial and microsomal preparations in different cell types [71, 86, 92 - 100].

Displacement studies using the classical androgens T and DHT and the steroid hormones 17β-estradiol (E2) and progesterone confirmed the specificity of the muscle intracellular AR binding sites. The androgen binding sites detected in total homogenates of C2C12 cells revealed affinity and specificity for androgen steroid competitors, as indicated by the displacement of [^{3}H] T by T and DHT, but not by E2 and progesterone. In accordance with the observed non classical subcellular location of AR binding entities, immunocytochemical assays with confocal microscopy by staining the cells with anti-AR antibody and Mitotracker red (MTT) or anti-Caveolin-1 antibody, confirmed the presence of immunoreactive AR entities in mitochondria and microsomes, respectively (Fig. **1**). Furthermore, the discontinuous sucrose density gradient fractionation revealed the presence of the androgen receptor in the low-density plasma membrane fragments, corresponding to lipid rafts and caveolae. Moreover, the non-classical localization of the androgen receptor in caveolae infer a physical interaction with Caveolin-1, an association that is lost after testosterone treatment, suggesting the androgen receptor translocation after T binding, from the membrane to some intracellular compartment. Thus, researchers provide evidence of the existence of extranuclear AR in the C2C12 cells [90]. The biochemical and immunological similarity between the reactive entities detected in mitochondria and membrane with the androgen receptor support the general idea of the presence of a subpopulation of AR with a non-classical localization, in skeletal muscle. The results presented in this work show an active role of the AR during the anti-apoptotic effect of testosterone in C2C12 skeletal muscle cells that might be performed at different levels: nuclear (in the genomic response), mitochondrial (in the intrinsic pathway) and microsomal (in the response mediated by membrane proteins).

Reproductive Tissues

The performance of immunoassays to detect the AR in human samples have demonstrated a strong staining reaction in all male reproductive tissues, producing basically the same pattern of staining in mouse, rat and human tissues. In general, the exclusive nuclear staining reaction corresponding to AR is detected in the secretory epithelial cells of the prostate, the seminal vesicle, and the epididymis [101, 102].

In prostatic alveoli, androgen receptor is located predominantly in the nuclei of epithelial cells. Tall columnar cells tended to exhibit intense positive nuclear staining whereas cuboidal cells showed less. Little stromal cells also show positive staining.

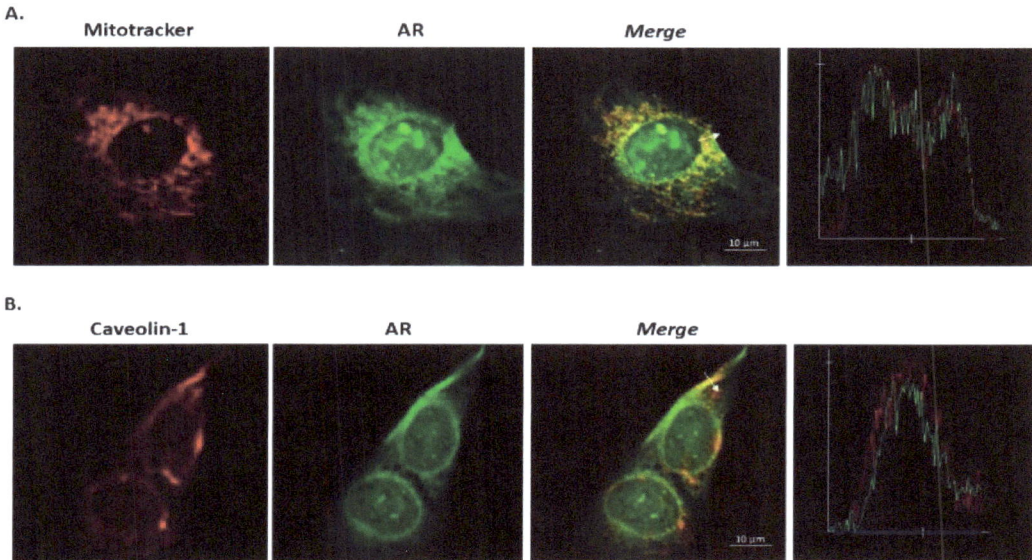

Fig. (1). Immunofluorescence location of the androgen receptor in mitochondria and membrane of C2C12 muscle cells. (A) Mitochondria of C2C12 cells were stained with Mitotracker (red fluorescence), and immunocytochemical assays were performed by confocal microscopy using anti-AR antibody (green fluorescence). *Merge* of both images is shown. Magnification: 63X. Fluorescence intensities were analyzed along the cells (white arrows in merge image), showing that the green fluorescence intensity profile matches up with the red intensity profile in many points of the graph, in agreement with the yellow signal observed. (B) C2C12 cells were double labeled using polyclonal antibodies anti-AR (green fluorescence) and anti-Caveolin-1 (red fluorescence). Magnification: 63X. Fluorescence intensity profile and merge of images are presented. Even though the yellow signal resulted from the merge of the images was weaker than that obtained for mitochondrial AR and the visualization of color yellow was hard to see, the intensity profiles recoded in green and red channels exposed the colocalization of both signals in some points of the graph (indicated with white arrows). Although the intensity profiles do not follow exactly each other and sometimes oscillate in contraphase, it was able to detect various points where green and red signals matched up. The difficulty to see the yellow signal in membranes, was possibly due to the huge amounts of Caveolin-1, the main component of caveolae membranes, respect to the androgen receptor, so that the green signal is dimmer than the red one and the overlap of colors still being seen as red.

In seminal vesicle, the pseudostratified secretory epithelium, which forms a highly convoluted gland, presents nuclear AR, some stromal cells contain the receptor too.

Glandular epithelial and stromal cells show little positive AR staining. The intensity is weaker than prostate and seminal vesicle.

In epididymis, the expression of AR in epithelial is highest in the proximal part, gradually decreases in the middle part, and increases again in the distal part. In stromal tissue, positive cells are observed in all parts of the epididymis. Besides, a considerable proportion of smooth muscle cells of these male accessory sex organs express the AR [101, 102].

In the testis, excepting spermatogenic cells, all other cells (Sertoli, peritubular and Leydig cells) show nuclear AR. However, the expression is weaker than in other male reproductive organs. Spermatogonia, spermatids, and spermatozoa do not seem to express the AR [72, 101].

In the stratified squamous epithelium of the penile prepuce of humans, the maximum expression of AR is in the basal cell layers, which gradually diminish in the more superficial layers.

Female reproductive organs also express the AR. The AR is located in the nuclei of stromal and stratified squamous cells in the basal and parabasal epithelial cell layers of the vagina. Nevertheless, intermediate and keratinized epithelial cell layers do not express the receptor. The smooth muscle layers show AR too, but the intensity is weaker than that in male reproductive organs. Myometrium and stromal cells of the uterus express the AR, but AR staining is absent or very weak in the endometrium and uterine glands. Except for the corpus luteum, major components of the ovarian parenchyma (follicles, stroma, interstitial tissue) do not express the AR [72].

Other Tissues

Even though in non-genital skin the epidermis and the hair follicles do not express the AR, human sebaceous glands, sweat glands, and ducts contain the receptor. Remarkably, the myoepithelial cells do not exhibit detectable AR expression, neither do dermal or subcutaneous mesenchymal cells. In mammary skin and associated sebaceous glands, the extend expression of AR in mammary acinar cells and inner ductal epithelial lining cells is variable and moderate, whereas myoepithelial do not express the AR [101].

In tissue constituents of the salivary glands, esophagus, stomach, colon, pancreas, bronchi, alveoli, and bladder, no presence of AR has been found. However, the nuclei of hepatic parenchymal cells (hepatocytes) in the liver have showed the expression of AR, whereas female tissue express AR to in a lesser extent. In the kidney, positive staining for AR is detected in the epithelial cells of convoluted tubules and in some of the parietal cells of Bowman's capsule.

Lymph nodes, thymus, thyroid gland, and parathyroid gland, have not revealed AR expression. The adrenal medulla is negative for AR staining, whereas the adrenal cortex contains positive cells. No difference between male and females in adrenal staining has been detected. The anterior pituitary gland also expresses the AR No difference has been observed between the male and female glands.

In peripheral neural tissue, including ganglion cells has not been found the AR [101]. Studies with rat cerebellum have revealed AR-positive cells in the Purkinje cell and granule cell layers [72].

BIOLOGICAL ACTIONS OF ANDROGEN RECEPTOR

Androgens are indispensable for the proliferation of the species and for establishment and maintenance of the quality-of-life of males by supporting sexual function and behaviour, muscle mass, strength and sense of well-being. Moreover, androgens orchestrate the development of male phenotype and function as important physiological regulators in many non-reproductive tissues of both genders. Many developmental events in males need androgens during a defined time window, whereas reproductive and non-reproductive functions are androgen-dependent during the course of the life. Testosterone, the main androgen, and its metabolite, DHT, are known to mediate their effects through binding to the androgen receptor. The classic function of testosterone includes androgenic and anabolic effects.

The AR is almost ubiquitously expressed and exerts a variety of roles in an enormous number of physiological and pathological processes. The comprehension of AR cell type- or tissue-specific physiological and pathophysiological roles employing *in vivo* mouse models offers valuable information in revealing AR roles in humans and ultimately helps to develop better therapies *via* targeting the androgen receptor or its downstream signaling molecules to battle androgens/AR-related diseases.

Reproductive Tissues

Reproductive tissues undergo enormous prenatal and pubertal developmental changes under androgenic control. Androgens are also required for their maintenance so that castration in adulthood results in regression of the male secondary sex tissues.

Prostate

The prostate gland is subjected to androgens for developing and maintaining its integrity. Congenital AR dysfunction or deficiency of 5α-reductase in genetic males causes slight or absent development of the prostate gland [103].

In animals as well, androgens are crucial for the integrity of the prostate gland. Within a week of castration, a rat prostate will involute resulting from epithelial cell apoptosis [104]. The growth of prostate in rat and dog is stimulated by DHT which appears to be the principal androgen in this tissue. In humans, evidence for the importance of DHT in prostatic growth may be derived from the absence of significant prostatic development in people with a 5α-reductase deficiency despite their high levels of T. The employment of a 5α-reductase inhibitor with exogenous testosterone inhibited prostate DHT development and caused regression of prostate size indicating that intraglandular DHT formation is essential for the anabolic effects of androgens on the prostate. Castration prevents prostatic development also in dogs and humans [1].

Testosterone is also a physiologic prostate cancer promoter. Research of eunuchoid persons suggested that prostates persist small and hypertrophy or prostate does not develop cancer [105]. Furthermore, animal models of prostate carcinogenesis need the existence of functional testes or exogenous T supplementation to support the development of prostatic cancer [106, 107]. Transgenic mice engineered for androgen receptor overexpression in the prostate, have a great turnover of prostatic epithelial cells and development prostatic intraepithelial neoplasia later in life [108]. Even though it is clear that androgens are indispensable for the development of prostate cancer, it has been hard to correlate relative levels of serum androgens with prostate cancer risk. This may be because the activity of the ligand-receptor may vary and play an equally important role in prostate cancer promotion.

Global ARKO mice models show no prostate development [109, 110] however, the prostate does actually develop in cell-specific prostate ARKO models. Prostrate epithelial ARKO (PEARKO) mice, in which the genomic activity of the androgen receptor is blocked in the epithelium of prostate, epididymis and vas deferens, shows decreased anterior and dorsolateral lobe weight, as well as augmented epithelial proliferation and abnormal epithelial clustering, mainly in the anterior lobe [111]. There is clear evidence in the PEARKO model of augmented sensitivity to estradiol, due to an increase of ERα expression in the dorsolateral prostate epithelial cells, suggesting that this receptor expression in prostate epithelial cells is modulated by local epithelial AR-dependent mechanisms [112]. Pes-ARKO mice, resulting in AR deletion in the ventral prostate and the dorsolateral prostate, show a phenotype of augmented ventral prostate size with increased epithelial proliferation in the ventral prostate and dorsolateral prostate [113]. The authors postulate that this model supports the hypothesis that androgen receptor modulates growth by suppressing epithelial proliferation. The peritubular myoid-specific ARKO (PTM-ARKO), resulting in AR deletion in smooth muscle cells in all prostate lobes, shows reduced prostate weight, hyperplasia, inflammation and increased estradiol sensitivity [114]. In contrast, the smooth muscle-ARKO, which has highest expression in the anterior prostate, has no change in gross appearance of the prostate but has loss of in folding structures into the lumen (mainly in the anterior prostate), reduced epithelial proliferation and decreased levels of insulin-like growth factor-1 (IGF1) expression [115]. A double stromal fibromuscular-specific ARKO (dARKO) mice which delete the AR in both stromal and smooth muscle cells demonstrated reduced prostate size due to reduced anterior prostate size and abnormal branching morphogenesis, with partial loss of glandular infolding structure, decreased proliferation and augmented apoptosis of the epithelium in the anterior prostate [116].

Role of the AR in Prostate Cancer

Prostate cancer (PC) is the most frequent solid organ cancer in men and the main cause of cancer death. Since Huggins and Hodges firstly demonstrated the sensitivity of PC to androgen deprivation, it has been clear that PC is dependent on AR activation for growth and survival [117]. Androgen deprivation therapy (ADT) is the mainstay of therapy in advanced PC, and also improves prognosis in appropriately-selected men with high-risk localized prostate cancer [118]. ADT is not curative, and after 1–2 years of ADT, clinical progression takes place [119]. Of significance, in most cases ADT-resistant prostate cancer remains androgen-dependent [119]. In spite of the very low levels of circulating T in serum, AR signaling is maintained by multiple mechanisms including activating AR mutations or truncations, AR amplification or overexpression that result in increased protein expression, changes in AR cofactor balance and extragonadal androgen production, including in the tumor tissue itself [120].

Generally, the *AR* gene is normal and expressed in primary PC. Nevertheless, after ADT, some *AR* gene alterations are found. These alterations conduce to an augmented responsiveness of the AR to low levels of circulating androgens and also to the ability of the AR to recognize a widened spectrum of ligands as effective agonists of the androgen receptor action. It might be possible thus, that AR signaling pathway is preserved in advanced PC that progresses after first-line androgen ablative therapy. Prostate cancer cells also respond to ADT by the amplification of *AR* gene copy number in patients who experience disease recurrence, which could reflect an adaptation of the cancer cells to castrate levels of circulating androgens. Remarkably, amplification of the *AR* gene is an adaptive response to high-dose antiandrogen monotherapy too, supporting the importance of augmented AR signaling for PC cells [121].

Hormonal deprivation also seems to conduct to selective mutations in the AR gene that modify its response to antiandrogens and extend the spectrum of ligand agonists. Some investigators have detected very rare AR mutations in PC tissue [122 - 125] in patients with primary PC, with higher frequency in patients with advanced disease, suggesting that mutations in the androgen receptor take place before hormonal ablative therapy and play a role in the progression of prostate tumor. AR mutations in PC are grouped in three areas of the receptor. In the LBD, mutations alter the ligand-binding pocket and open the spectrum of AR agonists to an extensive range of steroid hormones and pharmaceutical antiandrogens [126, 127]. AR mutations that alter the ligand-binding pocket do not, except for a single site, overlap at all with binding pocket mutations that cause androgen insensitivity [128]. Since AR mutations in PC are selected to increase AR activity and androgen insensitivity mutations reduce activity, this mutual exclusion is not unexpected. A second group is the area that alters binding of p160 coactivator molecules and the AR NTD [129]. AR activity is altered in PC, as well, by variations in AR coactivators. Hormone-resistant PC cells usually show AR overexpression and overexpression of coactivator molecules essential for androgen receptor signaling [130]. Mutations are also detected in the hinge region that borders the DBD and the LBD [128, 131]. The hinge region seems to be targeted because it alters AR interactions with corepressors and thus reduces the efficiency of antiandrogens and may explain the sensitization of AR to ligand interactions in late-stage PC [131]. Just as steroid hormone receptors initiate transcriptional signals that have to be amplified by coactivators, the signals can be silenced by corepressors. The hinge area of the androgen receptor between the DBD and LBD is often altered by mutations in PC that possibly mediates interactions with other proteins, perhaps corepressors. Besides, deletion of the hinge area amino acids 628 to 646 leads in noteworthy activation of the androgen receptor and marked enhancement of LXXLL-dependent ligand-dependent coactivation [132]. Remarkably, this area is basically never affected by mutations related to androgen insensitivity. This is in accordance with the idea that it is principally a site for corepressor binding and thus an important target for mutations in PC.

Epididymis

Androgen dependency of the other accessory sex organs such as epididymis, vas deferens, and seminal vesicles is also well known. Castration conduces in rapid regression of the secretory epithelium of the epididymis along with depletion of specific androgen receptors and reduction in 5α-reductase activity being reversed most of these alterations by an adequate androgen replacement. Exogenous androgen treatment does not completely restore epididymal secretory function [1].

It is possible that some androgen-dependent functions are regulated *via* seminiferous tubular fluid, which delivers large amounts of T and DHT to the initial segment of epididymis. Such high tubular androgen concentrations may be difficult to achieve by exogenous administration. The process of sperm maturation in the rat epididymis is dependent, at least in part, on specific secretory epididymal proteins, glycoprotein C and glycoprotein DE. The regulation of these specific proteins by androgens appears to occur both at synthesis level and at N-glycosylation [1].

Seminal Vesicle

The structural and functional integrity of the seminal vesicles, particularly the epithelial component, is dependent on androgens. Testosterone promotes the synthesis of the major secretory proteins by regulating their mRNAs. It has been demonstrated that administration of T but not DHT, to adult castrated male guinea pigs returned *in vitro* protein synthesis, cytoplasmic protein content, and wet weight to control levels in isolated seminal vesicle epithelium. Injection of diethylstilbestrol had an inhibitory effect on these parameters, indicating an antiandrogenic effect of estrogens in this tissue. The smooth muscle of seminal vesicle in guinea pig is sensitive to both androgens and estrogens, and the androgens appear to have a permissive role in estrogenic action on seminal vesicles. Thus, estrogens and androgens have a synergistic role in the maintenance or restoration of the smooth muscle component of seminal vesicles, whereas in the epithelium estrogens appear to be antiandrogenic [1].

The inhibition of prostaglandin synthesis suppresses the proliferative response of the seminal vesicle to T but not to DHT suggesting the involvement of prostaglandins in the mitogenic effect of T on the seminal vesicles as well as that T and DHT may have different mechanisms of action in this gland [1].

Penis

Androgens are essential for normal organogenesis of the penis, and castration causes mild involution. In rats, the penile spines, cornified structures projecting from the surface of the glans penis, are androgen dependent. Castration causes a decrease in their number and size while androgen treatment restores normal morphology. These structures function as tactile sense organs and influence sexual behavior. In addition to the central nervous system (CNS) effects, androgenic stimulation of sexual behavior in rats is in part due to peripheral effects. This peripheral component must be a minor one because DHT, a potent peripheral androgen that is presumably devoid of any CNS effect on libido, does not restore sexual activity in castrated male rats [1].

Testes

The effect of androgens on the testes is of special interest because of its dual function as the major source of androgens, and of gametogenesis. The identification of specific AR in rat Leydig cells suggests an autocrine function of androgens in these cells. Testosterone is required for normal spermatogenesis in mammals. Testosterone alone can stimulate spermatogenesis in hypophysectomized rats, monkeys, and in stalk sectioned rhesus monkeys in the absence of gonadotropins and in the face of high prolactin and estradiol levels. In hypogonadal men, similar findings have been reported but the interpretation of these reports is complicated by the fact that measurable endogenous gonadotropins were found in these patients. It has been suggested that the androgenic action on spermatogenesis is mediated by the Sertoli cells, which possess androgen receptors and show an appropriate temporal relationship between nuclear accumulation of androgen and stimulation of RNA polymerase II activity [1].

Central Nervous System

Since antiquity, humans have linked testosterone levels with behavior. Now, it is within the CNS that the relative roles of T, DHT, and estrogens in producing the varying effects of testosterone administration have been characterized. In 1959 it was shown that androgens acted directly on the brain. It was suggested that androgens, during early development, act to organize neural pathways responsible for male behaviors, whereas in adults, androgens act on differentiated pathways to activate previously organized behaviors [1].

AR in the brain shows similar physicochemical properties to prostate AR, recognizing both testosterone and DHT. In primates and rodents, AR is concentrated in the pituitary, hypothalamus, preoptic area, septum, and amygdala, with the lowest concentrations being in the cerebellum. Aromatization of testosterone to E2 varies from region to region also with the enzymatic activity being absent from the pituitary and cerebral cortex and present in the hippocampus, amygdala, and preoptic area. The 5α-reductase is distributed all over the CNS and, thus, does not appear to be a limiting factor in androgen occupancy of receptors [133]. However, 5α-reductase activity is higher perinatally than in the adulthood, suggesting a role for DHT in the neuronal organization [1].

Perhaps the best-studied aspects of sex steroid effects on the brain have been the reports of large anatomical sex dimorphisms. In 1973, Raisman and Field [134] showed that dendritic synapses in the rat preoptic area differed between sexes and could be manipulated by alterations in neonatal androgen levels. In 1976, Toran-Allerand [135] demonstrated that both androgens and estrogens enhance neurite outgrowth from hypothalamic explants in neonatal mice. Since then, five androgen-dependent anatomical areas have been clearly identified:

1. Greater size of brain regions involved in vocalization in male passerine birds than in females [136].
2. The sexually dimorphic nucleus of the preoptic area which is much larger in male rats [137].
3. Increased size in the middle portion of the medial amygdaloid nucleus in male rats [138].
4. Male/female differences in the neuronal number of motor neurons innervating the bulbocavernosus [139].
5. The smaller size of the superior cervical ganglion in females compared to male rats [133].

In many songbird species, the male sings whereas the female does not. Song has been shown to be a learned, complex, steroid-dependent behavior. The anatomical substrate of birdsong has been elucidated and consists of five forebrain nuclei as well as single nuclei in the thalamus, midbrain, and hindbrain. Several of these areas selectively take up androgens, and in where the female fails to sing even after administration of exogenous T, there are marked differences in the size of the brain nuclei involved in generating song [1].

Within the preoptic area, a cluster of intensely staining neurons, which has 3 to 7 times greater volume in the male rat, has been identified. This area has been named the sexually dimorphic nucleus of the preoptic area (SDN-POA). Castration of the newborn rat markedly decreases the volume of this nucleus, and the castration effects can be reversed by T administration [137]. Estrogens administered postnatally to females increase the volume of this nucleus, suggesting that it is the aromatization of testosterone to estrogen, that produces the masculinizing effects on the SDN-POA [139]. The amygdala is one of the main components of the extra hypothalamic gonadotropin-regulating system and plays a role in the modulation of a number of male behaviors. Electrophysiological studies have suggested that male rats have more synaptic connections projecting from the medial amygdala to the POA than do female rats or neonatally castrated rats. Nishizuka and Arai [138] have shown that the middle part of the medial amygdaloid nucleus of male rats has a significant increase in shaft synapses made on dendritic spines compared to female. Neonatal administration of T to females increases the number of synapses to male levels, and neonatal orchiectomy decreases the number to female levels. Thus, the medial portion of the amygdala appears to be another sexually dimorphic area of the brain [1].

The lumbar spinal cord nucleus of the bulbocavernosus (SNB) provides motor neurons to the bulbocavernosus and the levator ani which are attached to the penis in the male and are absent in the female rat. The female has only one-third the number of neurons in the SNB that the male has. Castration in adult male rats decreases both soma size and dendritic length of these neurons. These effects are reversed by testosterone treatment. Testosterone appears to control the neuronal number whereas DHT has a more potent effect on neuronal size [1].

Aromatase blockers inhibit testosterone-induced sexual behavior in castrated male rats, suggesting that aromatization of testosterone to estrogen plays an important role in the facilitation of male sexual behavior. DHT given alone has minimal effects on restoring sexual behavior in castrated animals. On the other hand, 17β-estradiol does not produce the complete pattern of male sexual behavior, whereas combinations of estradiol and DHT do fully restore sexual behavior [1].

The regulation of food intake and body weight is a complex process involving the integrative effects of multiple neurotransmitters and hormones. Adult male rats eat more and exercise less than adult females. Castration of adult male rats decreases food intake and locomotor activity. Low replacement doses of T restore food intake, whereas pharmacological doses further reduce food intake. DHT increases food intake. The decrease in food intake by high doses of T is blocked by progesterone, like the inhibitory effect of estradiol on feeding, suggesting that this effect is due to testosterone aromatization. On the other hand, estradiol stimulates physical activity in castrated males, while DHT does not. Antiestrogens attenuate T-induced activity, while the antiandrogen, cyproterone acetate, failed to inhibit androgen-induced running [1]. Thus, T and DHT appear to be the main hormones enhancing feeding in males, while aromatization to estrogens is necessary for the enhancement of physical activity in male rats.

In male N-ARKO mice, the androgen receptor is selectively knocked out in the brain, including important areas in the regulation of reproduction associated with neuroendocrine and behavioral functions, such as mating and aggression. Sexual motivation and performance, in N-ARKO mice, are fewer vigorous as well as aggressive behaviors, with only a little percentage of these mice which exhibit complete sexual behavior and offensive attacks. The erectile activity during copulating is also affected, while olfactory and neuronal activity are normal in N-ARKO mice in comparison with wild type. Therefore, central AR signaling is implicated in the regulation of male behaviors and modulation of neuroendocrine functions of gonadotropic and somatotropic axes [140]. Juntti *et al.* analyzed the

AR expression in the brain of developing mice, and observed that during testosterone surges in prenatal and neonatal stages, very little androgen receptor was expressed in the medial amygdala, bed nucleus of stria terminalis, and the hypothalamic medial preoptic region [141] which are central in the regulation of reproduction related neuroendocrine and behavioral functions, such as copulating and aggression. These investigators postulated that the androgen receptor is not implicated in the regulation of the organization and differentiation of neural circuits by controlling sexual and masculine behavior in mice. Instead, the AR is involved only in regulating the execution of these behaviors.

Human Sexual Behavior

The relevance of androgens to sexual function is controversial. In men, androgens appear to be necessary but not sufficient for usual libido. In healthy older men, high testosterone levels are related to greater sexual activity whereas, in younger men, such a relationship cannot be discerned. Lange *et al.* [142] suggested that latency to erection stimulated by erotic material correlates with testosterone levels [1].

Withdrawal of androgens in hypogonadal males conduces to a decline of libido 3-4 weeks later. Testosterone replacement restores sexual interest in a dose-dependent manner. Spontaneous erections are impaired in hypogonadal men, and Testosterone replacement improves the latency, frequency, and magnitude of the nocturnal penile tumescence response as well as the frequency of early morning erections [1].

Loss of AR in Central Nervous System Conduces to Hypothalamic Insulin and Leptin Resistance

The brain is an insulin target organ that exerts an important role in the modulation of energy balance and glucose homeostasis [143, 144]. Several studies have demonstrated that there is greater insulin sensitivity in the male than female CNS [145]. Higher levels of the AR are expressed in the hypothalamus of male than female mice and the AR was found to colocalize with the insulin receptor, suggesting that the AR of brain might be involved in the modulation of central insulin sensitivity in males [146].

The role of neuronal AR in metabolic regulation has been studied by Yu *et al.* by the generation of neuron-specific ARKO (N-ARKO) mice [146]. The supplementation of insulin in the fasting state diminished food intake in WT, but not in N-ARKO male mice, proposing that the latter had central insulin resistance.

Aged N-ARKO mice had greater body weight with augmented visceral fat, liver lipid deposition (related to an increased lipogenesis), and hepatic glucose production (associated with increased expression of gluconeogenesis genes, and glucose-6-phosphatase) than WT mice. Thus, ablation of neuronal AR resulted in hypothalamic insulin resistance that conduces to systemic insulin resistance, deregulation of glucose homeostasis and lipid metabolism, and visceral obesity [147].

Liver

A requirement of the liver on testicular secretions was first supposed when Kochakian [148] reported a small decline in the rate of growth of liver in castrated mice. These morphological observations were supported with the result of substantial androgen dependence of certain hepatic enzymes such as fumarase and catalase in mouse [148] and aspartic-glutamic transaminase, alanine-glutamic transaminase, and *d*-amino acid oxidase in the rat. Castration leads to a reduction of enzyme activity and T administration returns it to normal levels [148]. However, the effect of androgens on hepatic enzyme activity, in particular, some steroid-metabolizing enzymes, can be inhibitory. Thus, castration increases 5α-reductase and several hydroxylase enzyme activities, and Testosterone treatment reverses these effects. Testosterone treatment of female animals suppresses the activity of the female-specific 15α-hydroxylase enzyme, while castration of the male does not influence enzyme activity. The male-specific 16α-hydroxylase enzyme is neonatally "imprinted" and induced at puberty. Thus, neonatal androgens affect enzyme level and responsiveness in adulthood [1].

Although it is possible that some effects of androgen on the liver are mediated *via* AR, it appears that most of the *in vivo* effects of androgens on steroid metabolism are indirect and involve the interaction of androgens with other hormonal stimuli. Hypophysectomy abolishes the sex differences in hepatic steroid metabolism. Of the pituitary hormones tested, only GH administration results in an increase of 5α-reductase activity and a decrease in 6-hydroxylase and 16α-hydroxylase activities, suggesting that GH or a related factor is responsible for the feminization of hepatic steroid metabolism [1].

Hepatic microsomal drug-metabolizing enzyme activities are greater in male than in female rats. Orchiectomy reduces the activities of these enzymes, and androgens stimulate them. In this system, androgens appear to have two distinct effects: a nonreceptor-mediated stimulant effect on microsomal protein content and total liver weight and receptor-dependent effects that control, in part, the cytochrome P450-dependent system. The response of this system to androgens is attenuated by the antiandrogen flutamide. In testicular feminized mutant rats, with an inherited defect of AR, the hepatic microsomal enzymes do not respond to T treatment. The response of individual microsomal enzymes to androgens can be variable. The responses of the microsomal ethanol oxidizing system and microsomal steroid sulfatase are concordant in the male. In the intact female, T treatment induces microsomal ethanol oxidizing system activity but produces a decline in sulfatase activity. In the castrated female, T treatment did not alter steroid sulfatase activity [1]. These observations show that androgen-dependent control of hepatic steroid sulfatase is different from that of other microsomal enzymes, in that it can be modulated by estrogen or other ovarian factors.

Androgens can sometimes be considered antiestrogenic or estrogenic. It is of interest that oral 17-alkyl-androgen therapy has been implicated in causing hepatic lesions similar to those caused by estrogen employment, being comparable to the morphological changes of the liver induced by these steroids. Although the liver clearly shows sexual dimorphism, it appears that the majority of these differences is due to indirect effects of sex steroids on the production of pituitary hormones. In the majority of cases, the Testosterone effects appear to be stimulatory and the estrogenic effects inhibitory suggesting a primary effect of Testosterone rather than aromatization to estrogen. Although androgen-binding proteins have been identified in the liver, they appear to be different from the classical AR and may be more involved in steroid metabolism than in gene regulation [1].

Lipids

It is of relevance the effect of androgens on different lipoproteins because of the higher prevalence of atherosclerosis and shorter life span of men than women. In rat hepatocyte cultures, testosterone has a concentration-dependent biphasic effect on triglyceride synthesis: with lower doses being inhibitory and higher doses losing this inhibitory effect. Epidemiological studies have shown that high-density lipoprotein cholesterol (HDL-C) is lower in men and that triglyceride levels are higher compared to premenopausal women. Cross-sectional and longitudinal studies indicate that, during adolescence, HDL-C decreases whereas low-density lipoprotein cholesterol (LDL-C) and triglycerides increase in males. Testosterone administration at high concentrations causes a reduction of plasma HDL-C levels. On the other hand, inhibition of testosterone production with a GnRH analog in normal men produces a reduction of LDL-C and an increase in HDL-C concentrations. A carefully controlled study in healthy people concluded that Testosterone has a strong negative association with HDL in men and women when controlled for other variables [1].

The androgen effect could be due to enhanced catabolism of HDL. Sexual dimorphism has been reported in two enzymes of HDL metabolism. Lipoprotein lipase is greater in women than in men, and hepatic endothelial triglyceride lipase is greater in men and is induced by androgens and suppressed by estrogens. This is the enzyme primarily responsible for the clearance of HDL. Oral anabolic androgenic agents in women oppose the stimulatory effects of estrogens on the formation of HDL and very low-density lipoproteins and enhance the HDL clearance by the hepatic triglyceride lipase. The observation that boys with the highest estradiol levels had the greatest fall in HDL-C and the highest ratio of LDL-C to HDL-C with increasing T levels during puberty suggests that at least, part of the androgen effects are mediated by estrogens [1].

Loss of the AR in the Liver Conduces to Hepatic Insulin Resistance

The liver exerts an important role in regulating glucose metabolism and glucose homeostasis. Chronic increment of hepatic glucose production is the main contributor to hyperglycemia in diabetes mellitus [149, 150]. The liver also plays an important role in regulating lipid metabolism. Dysregulation of hepatic fatty acid synthesis and oxidation lead to hepatic steatosis (fatty liver) [151], which contributes to the development of hepatic insulin resistance with hepatic glucose overproduction [152].

A specific hepatic-ARKO mouse model was generated in order to study the role of AR in glucose and lipid metabolism. With a normal diet, Hep-ARKO mice showed a normal growth curve in comparison to WT, but male, and not female, hep-ARKO mice developed liver steatosis in the elderly. Upon feeding with a high-fat diet, male hep-ARKO mice gained weight and become more obese than WT mice. Obese male hep-ARKO mice developed more severe liver steatosis than male WT mice and hyperlipidemia, indicating dysregulation of lipid metabolism with a reduction in fatty acid oxidation and an increase in the synthesize fatty acid*s,* suggesting that it could be due to reduced PPAR-γ and increased SREBP1c expression levels, and alteration of their target genes involved in these metabolic events. Therefore, obese male hep-ARKO mice developed hepatic insulin resistance through the diminution in PI3K activity, the increase in phosphotyrosyl phosphatase 1B (PTP1B) levels and phosphoenol-pyruvate carboxykinase (PEPCK) expression in the liver, leading to the increase of hepatic glucose production. These observations suggested that the hepatic androgen receptor plays a crucial role in the modulation of hepatic glucose, lipid metabolism, and insulin sensitivity [153].

Kidney

The stimulatory effect of androgens on kidney weight is accompanied by modifications in the epithelial cells of convoluted tubules, parietal cells of Bowmen's capsule [148], and glomerular volume occurring concomitantly with increased RNA and protein synthesis. In ovariectomized female rats, treatment with estradiol, DHT, 19-nortestosterone, or methyltrienolone results in an increase in renal mass. Simultaneous administration of flutamide with methyltrienolone or DHT blocks the renotrophic effect of these androgens, indicating that AR is mediating these effects. The simultaneous administration of maximum doses of estrogens and androgens to male or female castrates leads to an increase in renal mass exceeding the normal female/male range, suggesting independent action of these steroids.

It is possible that some of the effects of androgens on the kidney could be mediated *via* progestogen receptors. It is noteworthy that the classification of progestins as androgenic, synandrogenic, or antiandrogenic is based on their ability to mimic, potentiate, or block the effects of androgens on the rate of synthesis of β-glucuronidase in renal proximal tubules and the rate of excretion of this and other lysosomal enzymes into urine [1].

Similar to androgen effects on liver, some androgenic responsiveness of the kidney is mediated by the hypothalamic pituitary unit. Hypophysectomy reduces ornithine decarboxylase (ODC) activity in the rat kidney but has no effect on ODC activity of mouse kidney. Similarly, androgenic control of mouse renal alcohol dehydrogenase and D-amino oxidase activities is independent of the pituitary gland, while β-glucuronidase activity is partially dependent on pituitary hormones. Androgens may also stimulate endocytosis as well as hexose and amino acid transport in the mouse kidney cortex. This effect has been related to the androgenic stimulation of extracellular calcium influx and mobilization of intracellular calcium [1].

There is no apparent human analogy to the renotropic effect of androgens in rodents. Although anabolic steroids have been used in humans with acute and chronic renal failure, there is no evidence that the modest reduction in azotemia is due to improved glomerular filtration. Administration of anabolic steroids to elderly, non-uremic subjects results in significant increases in glomerular filtration rate, renal plasma flow, and excretory rate of para-aminohippurate (TmPAH). With the exception of the last parameter, these alterations are transient and attributed to the androgen-induced hypervolemia.

Summarizing, testosterone has a direct effect on mouse kidney growth and enzyme activity, independent of aromatization. Some of the effects of androgens on the kidney are blunted by hypophysectomy, suggesting that they may be indirectly mediated through the hypothalamic-pituitary axis. With the exception of stimulating erythropoietin secretion, there does not seem to be any important effects on human kidney [1].

Hematopoietic System

Over several years, numerous studies have indicated that androgen may be beneficial in the treatment of a variety of primary anemias and bone marrow (BM) failures. The effect of androgens on erythropoiesis has been inferred from the general observation that the circulating hemoglobin concentration increases in boys at puberty concomitantly with the increase in serum testosterone. In hypogonadal males, the hemoglobin level is reduced, and testosterone replacement therapy restores the levels. Women receiving androgens for treatment of arthritis have significant increases in red cell mass [1].

The erythropoietic action of androgens involves the stimulation of erythropoietin production *via* receptor mediated transcriptional control and a direct effect on BM. Androgen administration to polycythemic or starved rats and mice results in increased circulating erythropoietin levels. Although the kidney is the major source of erythropoietin, extra renal sources are also involved since androgens can produce an erythroid response in nephric patients, and severe hemorrhage induces erythropoietin secretion in nephrectomized rats.

Androgens also directly stimulate stem cell of the BM leading to the proliferation of other blood cell lines. *In vitro* BM culture studies have shown that testosterone can induce erythropoiesis. Testosterone enhances the two erythroid stimulators, the colony-forming unit (CFU-E) and the burst-forming unit (BFU-E) [1]. In addition, the synthesis of heme (A) and globin is stimulated by androgens. The effects of androgens on granulopoiesis and platelet production were demonstrated in the accelerated recovery of the white blood cell and platelet count after radiation or chemotherapy. Similarly, rats pretreated with T have less depression of hemoglobin, white blood cell, and platelet counts from ^{32}P therapy.

Male rats have a greater predisposition to arterial thrombosis formation and platelet aggregation than females. However, the administration of testosterone to white carneau pigeons does not alter collagen, ADP, or arachidonic acid-induced aggregation or the synthesis of prostaglandins in thrombocytes [1].

Although estrogens have an inhibitory effect on erythropoiesis, the trophic effect of androgens does not appear to be an antiestrogenic effect because estrogens do not alter the proliferation of erythroid precursors in culture and the *in vivo* stem cell-suppressive effect of estrogen is not observed in human tissue. In a study comparing the erythropoietic, androgenic, and myogenic activities of 29 different T derivatives [154], 1-dehydro-17-methyltestosterone, an anabolic steroid, demonstrated erythropoietic and myogenic activities without substantially stimulating the development of ventral prostate and seminal vesicles. Similarly studies in polycythemic mice demonstrated that the administration of α-derivatives of androgenic steroids enhances red cell production to the same degree regardless of whether they were predominantly androgenic or anabolic [155], suggesting that the erythropoietic activity can be independent of the androgenic or anabolic effects of the steroids. Thus, the action of androgens in the hematopoietic cells appears to be different from that in the reproductive tissues.

It has been well established that androgens induce erythropoiesis. The role of androgens became evident after the initial observations of men having higher red cell mass than women and the induction of hemoglobin synthesis in response to androgens in animal experiments.

Animal experiments in the 20th century exhibited that androgen administration to various animals lead to an increment in the levels of hemoglobin and BM activity [156]. Steinglass *et al.* demonstrated the relevant role of androgens in hematopoiesis by showing that castrated rats exhibit an important decrease in hematopoietic activity [157]. Additionally, the supplementation of androgens to such animals lead to an improvement in red blood cell amounts [158]. First human studies revealed that adult men have higher hemoglobin and red cell amount than adult women [159, 160]. Interestingly, this difference in blood parameters between men and women is not evident before puberty and only after the pubertal spurt, men gain an advantage over women in their erythropoietic machinery [161]. Although in women who do not have any menstrual blood loss, men still had significantly higher red cell amount. In accordance, men who have suffered castration develop anemia and supplementation of androgens reestablishes their red cell mass [156, 157]. The role of androgens in erythropoiesis is further supported by the observation of augmented red cell mass in women with hyperandrogenism [162]. Likewise, women with breast cancer receiving huge doses of androgens also show an increase in red blood cell mass [163]. These observations demonstrate that androgens play a noteworthy role in the induction of erythropoietic machinery.

Studies propose that androgens act both directly and indirectly to induce erythropoiesis. Studies in animal and human showed that androgen supplementation increases the synthesis and secretion of erythropoietin (Epo). Administration of androgens to female rats leads to hypertrophy of renal tissue and an increment in Epo secretion [164]. Testosterone binds specifically to renal cell cytoplasmic receptors and is transported to the nuclei where it enhances RNA polymerase activity within a few hours, leading to an increment in renal mass [165]. This conduces to the increase of Epo synthesis and secretion. Androgens also induce Epo synthesis in humans. Elevated concentrations of Epo have been observed in the urine of men on androgens. Medlinsky *et al.* further analyzed the idea that augmented Epo synthesis is the central mechanism by which androgens induce erythropoiesis [166]. They employed a testosterone antagonist in mice, which provokes an important diminution in erythropoiesis.

Androgens stimulate the CFU for erythroid in the bone marrow. After entry into the BM, androgens are reduced to 17-keto-steroids, binding then to the nucleus and increasing the synthesis of mRNA. This mRNA generation in uncommitted BM cells leads to their conversion to Epo-responsive cells. From then on, Epo is necessary to further enhance the maturation of these cells into an erythroid cell line. Moriyama and Fisher have demonstrated that a combination of T and Epo have a synergistic effect on erythropoiesis [167]. Testosterone firstly acts on the marrow cells to make them Epo-responsive and then Epo differentiates these pluripotent cells into an erythroid cell line. Testosterone is also known to induce iron (Fe) incorporation into erythrocytes [168].

Erythrocytes express ARs, and androgens circulate freely into and out of the erythrocytes [169]. After the entrance into the cells, androgens increase the uptake of glucose, leading in glycolysis [170]. This intracellular phosphorylation results in the formation of high energy phosphate bonds conducing to DNA transcription and the synthesis of mRNA in the erythroid cells which enhance the synthesis of red blood cells [171]. Erythropoiesis is detected after 6-12 hours of glucose utilization [170]. This increment of glycolysis is observed after supplementation with all types of androgens [172].

In vitro studies have demonstrated that the synthesis of hemoglobin by human marrow cells is augmented by the administration of androgens [173]. Similar results have been observed in several animal cell cultures [174]. Nevertheless, the increase in hemoglobin synthesis by androgens seems to have a minor contribution in erythropoiesis in comparison with other mechanisms.

Immune System

Autoimmunity

Systemic lupus erythematosus (SLE), a hallmark systemic autoimmune disease, has a female to male ratio of 9:1. Some of the first researches focusing on the role of androgen in autoimmunity were performed on mice in the early 1970's and showed an important protective role of T during lupus-like disease development [175, 176]. Mice were protected from disease development by the presence of T while castration lead to the development of lupus-like disease symptoms similar to those observed in female mice. Of relevance, female mice with severe disease exhibited a reduction in disease severity and prolonged survival upon treatment with T. The protective effect of T is not only limited to SLE but also it has been shown to exert a protective role in other autoimmune diseases such as multiple sclerosis, rheumatoid arthritis and arthritis [177, 178]. Therefore, T seems to have a suppressive action on the immune system and play a central role in prevention of autoimmunity.

Cancer

In the case of autoimmunity the immune system is commonly believed to be hyper-activated, whereas in cancer the immune system fails to support the required immune response to battle cancerous cells. It was shown that males are more predisposed to develop cancer than females [179, 180]. In a follicular thyroid cancer research, it was demonstrated that testosterone promotes tumor progression by suppressing tumor immunity *via* inhibiting tumor-infiltrating CD8[+] T cells and M1 macrophage [181]. Likewise, in an early colonic cancer model, castration significantly protected male rats from developing colonic adenomas [182]. Testosterone appears to have a suppressive role in the immune response to cancer and consequently may act as a potential promoter of tumor growth.

Neutrophil Development and Acute Infection

It was found that androgens induce proliferation of committed erythrocytic and granulocytic precursors *in vitro* [183, 184], and induce the recuperation of leukocytes after chemotherapy or radiation for cancer in rats [185]. Observations after androgen treatments of aplastic anemia [186] and during therapy with myelosuppressive agents [187], propose that androgen might induce neutrophil production. The polycystic ovary syndrome (PCOS) patients with elevated levels of androgen in serum, also show higher neutrophil amount [188]. These observations point to a possible participation of androgen/AR signaling in granulopoiesis.

Studies in AR knockout mice (ARKO) have provided the knowledge about the role of the androgen receptor in neutrophil production and function. These mice have less neutrophil amounts in the blood and BM and were more prone to acute bacterial infection. A reduction in myelocytes/metamyelocytes and a marked diminish of mature neutrophils were found in BM cells of ARKO mice in comparison with those of wild type (WT) mice. These findings point to that the deficiency in granulopoiesis of ARKO mice takes place during the transition between the proliferation of precursors (myeloblasts, promyelocytes, and myelocytes) and maturation of neutrophils (metamyelocytes, band forms, and neutrophils), conducing to a difference in terminal differentiation of neutrophils.

Further analyses have showed that BM granulocytes from ARKO mice have lower levels of mitotic cell division and are more prone to apoptosis in comparison with those from WT mice. Moreover, neutrophil production in response to granulocyte colony-stimulating factor (G-CSF) administration was decreased in ARKO mice in comparison with WT mice. Bone marrow granulocytes of WT mice treated with AR-siRNA decrease ERK1/2 activation and cell proliferation in response to G-CSF, while retroviral expression of the androgen receptor in ARKO granulocytes promotes ERK1/2 activation and cell proliferation [189]. Besides, ablation of the AR in granulocytes results in impairment of G-CSF signaling to influence neutrophil precursor proliferation and differentiation.

Of interest, the androgen receptor in neutrophils appeared to play an important role in the innate immune system to defend against microbial infection based on the observation that ARKO mice show augmented death rates in comparison with WT mice upon pathogenic bacterial and septic challenges [147, 189].

Innate Immunity

The innate immune system develops in the BM from common myeloid progenitors (CMPs). Due to the presence of the androgen receptor in hematopoietic progenitors, there is a reason to believe that T may play a significant role in shaping the immune cell repertoire even before the cells leaving the BM. Neutrophils and monocytes are cells of myeloid origin and are the first to respond to pathogenic infections. Mature neutrophils and monocytes are released in the periphery where they circulate in the blood until being called to sites of infection or injury.

It has been demonstrated the existence of immunosuppressive neutrophils, resembling granulocytic myeloid-derived suppressor cells (MDSCs), in protected male lupus-prone mice [190]. These investigations demonstrated that not only male mice contained elevated levels of these immunosuppressive regulatory neutrophils, but also the cells were modulated by T and protected against the development of lupus-like diseases [190]. Supporting a linking between T and neutrophils, granulopoiesis is significantly decreased in AR-deficient non-autoimmune C57Bl/6 mice [189]. More precisely, neutrophils from AR-deficient mice are less reactive to G-CSF-induced proliferation, more prone to apoptosis, and defective in their response to migratory signals *in vitro*. Therefore, neutrophil homeostasis is subjected to functional AR expression. Of relevance, expression of the androgen receptor by myeloid progenitors is in itself modulated by T [189].

Monocyte/Macrophage Production and Inflammation

Macrophages constitute a central part of the innate immune system and are greatly pleiotropic in function. Depending on the localization and origin of the macrophages, they can exert multiple functions ranging from tissue homeostasis, remodeling, clearance of apoptotic bodies, capturing antigens, and secreting many cytokines shaping the immune response [191, 192]. The early myeloid progenitors of the BM give rise to monocyte/macrophage-restricted progeny under the influence of granulocyte-macrophage colony-stimulating factor (GM-CSF) and macrophage colony-stimulating factor (M-CSF) and eventually develop into monocytes, which are then transported to the circulation. Upon stimulation, blood monocytes migrate to peripheral tissues, particularly the inflammatory sites, and differentiate into tissue macrophages [193].

The analysis of the peripheral blood mononuclear cells (PBMCs) has revealed that the monocyte population, especially in the inflammatory monocyte population is decreased in male ARKO mice in comparison with WT mice. However, the resident monocyte subset in PBMC was similar between WT and ARKO mice, suggesting that the AR might specifically enhance the inflammatory monocyte pool in the peripheral blood. The expression of a key chemokine expression receptor, CCR2, and TNFα in BM-derived macrophages from ARKO mice are decreased respect to WT mice. These findings, suggest that the androgen receptor is implicated in the regulation of inflammatory monocyte/macrophage production, migration, and function, although the mechanism through which the AR regulates these processes remains unclear [194].

B Lymphopoiesis

Several studies have enunciated that androgen/AR signaling might regulate the production of B lymphocytes and influence autoimmunity. For example, Viselli *et al.* saw that in castrated male mice had enlarged spleens with the expansion of B cells in the spleen and BM in comparison with intact mice [195]. The androgen administration resulted in a reversal of B cell expansion. Whereas castration of male mice leads to augmented lymphopoiesis and thus more B cells and B cell precursors (early pro-B cell and the late pro-B cell), ADT of castrated mice replenishes the numbers of B cells back to normal levels [196], suggesting that androgen/AR signaling regulates B lymphopoiesis [197]. Altuwaijri *et al.* observed that in mice with AR functional ablation presented an increase in population of B cells in BM and spleen [198], with a greater proportion of pre-B cells and inferior proportion of pro-B cells than WT BM cells. Changes of B precursor cells seems to be related to decreased apoptosis of B cells and augmented proliferation of B precursor cells in ARKO mice than WT mice. The resistance of AR-deficient B cells to apoptosis is accompanied by altered levels of many modulators of apoptosis. Since apoptosis exerts a relevant role during B cell development and tolerance [199], the increase in apoptosis in AR-deficient B cells would be likely to affect B cells' specificity and autoimmunity.

Given that both BM stromal cells and B cell progenitors express AR, chimeric mice in which the expression of the androgen receptor was limited to either stromal cells or lymphoid cells within the BM, were created to identify the primary cellular target of androgens in the BM [200]. These studies demonstrated that AR expression in stromal cells, but not lymphoid cells, is necessary for testosterone's inhibitory effect on B cell lymphopoiesis [200]. Of relevance, the negative effect of T is not directly as consequence of reduced IL-7 production by stromal cells, but rather *via* the induction of TGF-β secretion by BM stromal cells, resulting then in the reduction of IL-7 levels [200, 201]. Low levels of androgen or abnormally little expression of the androgen receptor correlates with an increased proneness to rheumatoid arthritis, which is present in males with low levels of androgens and patients with prostate cancer undergoing ADT, as well [198]. This is in part a consequence of augmented proliferation and reduced apoptosis of B lymphocytes, letting auto reactive B cells to expand into the periphery [198]. Finally, these studies point to a central role of androgens in the regulation of the development of B cells as well as the inhibition of the expansion of potentially auto reactive B cells in the periphery, conducing to autoimmune disease. Finally, even though castration of normal male mice also leads to the expansion of newly emigrated immature B cells within the periphery [198], it is important to note that B cells within the spleen do not express AR and thus, the effects induced by androgens are likely directed towards B cell lymphopoiesis [202].

Thymocyte Development and Peripheral T Cell Population

T lineage-committed hematopoietic progenitor cells formed in BM migrate to the thymus where they differentiate into mature thymocytes, and emigrate from the organ and home to many lymphoid tissues so that mature T lymphocytes can trigger an adaptive immune response.

Several studies have shown that androgen deprivation in males is related to enlargement or rejuvenation of the thymus [203 - 205], suggesting that androgen/ AR signaling is involved in modulating thymic function. Ablation of androgen by castration of male animals and a defect in androgen signaling, leads to reversion and re-expansion of the involuted thymus, accompanied with an increment in the number of cells due to more proliferating and fewer apoptotic thymocytes [203, 205]. The reversal of thymic atrophy post castration is age- independent and can be reverted with testosterone or DHT supplementation, resulting in the inhibition of thymic rejuvenation, reduced proliferation and incremented apoptosis of thymic T cells [204]. These results have also been proved in humans by demonstrating that restoration of testosterone in androgen-deficient men lead to the decreased thymic output of T cells [206]. Taken together these results show an inhibitory effect of androgens on the development of the thymus and the modulation of thymocytes. The response of the thymus to changes in androgen levels suggests that the effects are mediated by the androgen receptor.

Thymic involution leads to the decreased export of T cells into the periphery, and thus, a reduced peripheral T cell receptor (TCR) repertoire and function [207, 208]. Castration of male mice reveals an important increase in recent thymic emigrants, specifically CD44^{low} naïve CD4$^+$ and CD8$^+$ T cells, replenishing the peripheral T cell pool [203]. These studies demonstrate that androgens act to inhibit the number, and possibly the repertoire, of recent thymic emigrants entering the periphery. In conclusion, proliferation of splenocytes in response to TCR stimulation increased after castration of male mice in an age-independent way, proposing that androgens may alter the magnitude of T cell responses after immunization [203, 209].

Cardiovascular System

The role of androgens in the cardiovascular system in humans is complex and confusing. Human males have a considerably higher risk of cardiovascular disease than females, but in the elderly males with lower levels of testosterone have greater cardiovascular mortality than males with higher testosterone. Moreover, males with cardiac disease, mainly with ventricular failure, have lower levels of T. A recent trial of T administration in males provoked an increment in cardiovascular events [210]. Thus, the role of testosterone in cardiovascular system may be important, but remains poorly understood.

Cardiac Growth

Although it has been reported that there are no differences in systolic blood pressures and heart rates between wild type and ARKO mice, the size of heart or heart weight/body weight is smaller in ARKO than in WT mice. Cross-sectional analyses of the hearts indicate that also the volume and wall thickness of the left ventricles are smaller in ARKO in comparison with WT mice. To evaluate the role of the AR in cardiac growth, Ikeda *et al.* employed 25-week-old ARKO and WT mice, which were stimulated with or without Ang II during 2 weeks [211]. The importance of the AR in cardiac development and function was highlighted by the phenotype of ARKO mice. They had an important reduction in cardiac hypertrophy stimulated by Ang II and heart-to-body weight ratio in comparison with WT, events associated with lower activation of extracellular signal-regulated kinases (ERKs) 1/2 and ERK5. Moreover, impairment of left ventricle function and cardiac fibrosis induced by Ang II was reduced in ARKO mice [211].

Vascular Remodeling and Atherosclerosis

Vasoconstriction in response to KCl, mediated by voltage-gated calcium channels, and vasodilation in response to acetylcholine, which is endothelium-dependent, have been reported to be decreased in femoral arteries of testicular feminization mutation (Tfm) mice in comparison with WT, suggesting that androgen/AR signaling might be modulating arterial functions. Nettleship *et al.* showed that a high cholesterol diet during 4 weeks in Tfm and castrated male WT mice lead to a significantly higher fat deposition in the aortic root than in intact WT mice [212], suggesting that androgen/AR signaling might protect against atherosclerosis.

The effect of AR knock out on vascular remodeling was also evaluated by the employment of ARKO mice [213]. Despite nitric oxide (NO) availability, aortic endothelial NO synthase expression and phosphorylation, as well as Akt phosphorylation, were significantly decreased in ARKO mice. Upon Ang II stimulation, however, ARKO mice showed marked increases over WT mice in thickness and perivascular fibrosis of the aorta that were accompanied with augmented expressions of TGF-β1, collagen type I, collagen type III, and nicotinamide adenine dinucleotide phosphate oxidase component genes, as well as incremented superoxide production and lipid peroxidation. Besides, phosphorylation of c-Jun N-terminal kinase (JNK) and Smad2/3 was increased in ARKO over WT mice after Ang II stimulation. These observations lead to conclude that the androgen/AR system is necessary to the preservation of NO bioavailability through activation of the Akt endothelial NO synthase system and has protective effects against Ang II-induced vascular remodeling by modulating oxidative stress, JNK signaling, and the TGF-β1/phospho-Smad signal cascade.

Studies of atherosclerosis in male LDL receptor deficient (LDLR-/-) mice with WT AR, ARKO, smooth muscle cell-specific ARKO (SM-ARKO), myeloid cell specific ARKO (Myl-ARKO), or endothelium-specific ARKO (End-ARKO) have demonstrated that the androgen receptor may exert a positive role in promoting atherosclerosis in selective cell types [214]. The androgen receptor in non-myeloid cells like smooth muscle and endothelial cells might not play any significant role in atherosclerosis, while AR in myeloid cells has a positive role in promoting atherosclerosis. Due to very low T concentrations in ARKO mice, it is unclear whether ERα or the AR in extra aortic cells might protect against atherosclerosis by regulating cholesterol and lipid metabolism. The underlying mechanisms of these AR roles remain to be elucidated [147].

Bone

During aging fractures are a major and classical health problem. The role of sex hormones in the pathogenesis of osteoporosis has been widely studied in women, but not too much in men. The fact that the decrease in T levels in men does not occur as quickly as the decrease in estrogens in women after menopause, contributes to this disparity. Two mechanisms exerted by androgens are known to contribute to reduce the risk of fracture in aged men:

1. Androgens exert an important role in determining highest bone mass or maximum bone density as they are vital for skeletal growth and bone accrual in puberty [215].
2. Androgens maintain bone in post-pubertal males [216] defining both the size and strength of adult bone [217].

The relevance of androgen effects on bone mediated directly by the androgen receptor or indirectly following aromatization to estrogens and through the estrogen receptor remains to be elucidated. The anabolic effects of androgens, in conjunction with their potential anticatabolic action, makes androgen physiology a crucial candidate for understanding the bone fragility of aged men.

In ARKO mice males, it has been observed bones with reduced size, thickness and volume in comparison with controls [218, 219], indicating the requirement of the AR in these key actions. Yeh *et al.* showed that male ARKO mice had not only lower cancellous bone volumes than male and female WT mice, but also higher osteoclast numbers, mineral deposition, and bone formation rates in the femoral metaphysis [220]. The osteopenia in ARKO mice was endorsed to a higher bone resorption rate than bone formation rate. Tsai *et al.* observed that male ARKO mice, showed continuous loss of bone mass and deterioration of bone microarchitectures (reduction in trabecular number and augmented trabecular separation) in trabecular bones respect to WT mice [221]. MacLean *et al.* observed that male truncated AR^{-Ex3} mice exhibit reduced trabecular number and volume, bone mineralization surface, cortical bone thickness, periosteal, and medullary circumferences in comparison with WT male mice [218]. On the contrary side, female AR^{-Ex3} mice display only reduced trabecular number, periosteal, and medullary circumferences and augmented trabecular thickness in comparison with WT females. These observations suggest a significant role of AR in bone metabolism in both male and female mice.

It has been reported that when mesenchymal stem cells (MSCs) are encouraged to differentiate into osteoblasts, the expression of many genes associated to osteogenesis are reduced in cells isolated from ARKO mice in comparison with Wt mice. Therefore, the AR mediates the regulation of MSC proliferation and osteogenic differentiation [222].

There is substantial evidence proposing that the actions of androgens on bone are mediated, at least in part, by the AR present in osteoblasts and osteocytes [223, 224]. Transgenic mice that overexpress the androgen receptor precisely in proliferating osteoblasts, including those localized at the periosteum, have larger bones due to incremented periosteal mineral apposition [225]. On the opposite, overexpression of the androgen receptor in mineralizing osteoblasts does not have an effect on bone size [226]. However, both AR transgenic models have a common phenotype of augmented trabecular bone volume as a consequence of decreased bone turnover [225, 226]. Accordingly, the deletion of the AR specifically in osteoblasts and osteocytes from either the i) mature or ii) mineralization stage of osteoblast maturation [227] has the contrary effect of trabecular bone loss because of the augmented bone resorption [228, 229]. This is demonstrated by a decrease in trabecular bone volume and number, in the osteoblast-ARKO mice, compared to controls [228, 229]. Decrease in cortical bone and deregulation of the bone matrix synthesis and mineralization processes are also detected in mice that do not express the androgen receptor in mineralizing osteoblasts, identifying a relevant role of the AR in the regulation of bone resorption and mineralization, principally throughout pubertal growth when quick bone formation is needed. Moreover, the generation of osteocyte-specific ARKO mice reduced trabecular bone, but did not impair its response to mechanical loading. Expression of the AR in osteocytes enhance with age, but are decreased greater than 80% in osteocyte-specific ARKO (Ocy-ARKO) osteocytes in comparison with WT osteocytes [230]. Taken together, these results propose that androgen action *via* the AR in osteoblasts, is dependent on the stage of osteoblast maturation with AR activation in mineralizing osteoblasts and osteocytes, inhibiting bone resorption within cortical and maintaining the integrity of trabecular bone whereas activation of the AR in proliferating osteoblasts modulates an anabolic action on cortical bone at the periosteum.

Muscle

Skeletal Muscle

Androgens, through its actions on muscle mass and adipose tissue, is an essential determinant in body composition in male mammals, including humans. Testosterone is important for development and maintenance of muscle mass and strength [231, 232].

Testosterone administration enhances muscle mass in healthy young and old men, in healthy hypogonadal men and in other physiological or pathological conditions that take place with low levels of the hormone and decline the fat mass in eugonadal young and aged men [233]. Herbst and Bhasin postulated that androgens induce the commitment of mesenchymal pluripotent cells into the myogenic lineage and block adipogenesis through an AR-mediated pathway [234]. The increment of muscle size induced by T is linked with hypertrophy of muscle fibers and an important increase in the number of myocytes and satellite cells [235 - 237]. Given that numerous actin-interacting proteins function as AR coregulators, Ting and Chang postulated androgen/AR signaling plays significant roles during development and function of skeletal muscle [238]. Nevertheless, it is unclear at present how the AR in these cells acts to regulate skeletal muscle mass and strength.

Altuwaijri *et al.* showed that through differentiation of myoblasts into myotubes *in vitro*, expression of AR and myogenin increased, suggesting that AR signaling might be involved in muscle development [239]. However, when hind limb skeletal muscle from adult ARKO and WT male mice were compared, no differences in muscle morphology, desmin protein levels (a muscle-specific protein), or β-dystroglycan (a muscle-specific membrane protein), between the two genotypes, were detected, pointing to a normal muscle differentiation in ARKO and WT mice. On the contrary, the levels of myosin and troponin I specific for slow-twitch muscle fibers were diminished, while troponin T specific for fast-twitch muscle fibers was augmented in quadricep muscles of ARKO mice in comparison with WT mice. Thus, AR might exert a role in skeletal muscle development by influencing the balance of muscle fiber type in favor of slow twitch fibers through positive regulation of slow-twitch fiber-specific proteins and negative regulation of fast-twitch fiber-specific proteins [239].

Studies in rats have also shown that exogenous administration of testosterone results in a more rapid recovery of hind limb paralysis induced by sciatic nerve injury [240], as well as preventing apoptosis of muscle cells of the levator ani muscle in castrated rats [241]. With the employment of ARKO mice Callewaiert *et al.* showed that ablation of the AR reduced muscle mass [242], pointing to an important role of AR signaling in growth and/or maintenance of muscle mass. Ophoff *et al.* generated a myocyte-specific ARKO mouse (M-ARKO) [243] in which AR protein expression was decreased by 60-88% in comparison with muscles of WT mice. The residual AR expression was attributed to non-myocytic muscle cells. In comparison with WT mice, muscles of M-ARKO mice exhibited greater proportion of slow-twitch fibers and decreased fast-twitch muscle fibers. On the other hand, there were no differences in muscle strength and fatigue between the two genotypes. Thus, it was concluded that myocytic AR acts to preserve muscle mass and regulate muscle fiber type.

T treatment in fragile ageing men with low-borderline T levels, prevented the loss of limb strength and improved body composition [244]. Androgens are necessary to preserve muscle mass in men, with androgen withdrawal leading to muscle atrophy. Loss of muscle mass in adult men following androgen withdrawal has been proved in both normal men and men with PC undergoing ADT [245, 246]. The decay in T levels with age in males may be one mechanism causative of the age-related loss of muscle mass that takes place in ageing fragile men.

The loss of muscle mass and strength in the elderly, also referred to as sarcopenia, is a greatly predominant condition during aging and predicts numerous adverse outcomes, including disability, institutionalization, and mortality. Age-related muscle loss is a consequence of the decrease in size and number of muscle fibers [247], possibly because of multifactorial process that includes absence of physical activity, decreased nutritional intake, oxidative stress, and hormonal changes [248, 249]. Sarcopenia has been related to a deficit of sex hormones as the levels of E2 and/or T decline in the elderly. Even though the exact mechanisms causal of sarcopenia are far from being clarified, collecting evidence points to an age-related acceleration of myocyte loss *via* apoptosis as a crucial mechanism responsible for the impairment of muscle performance [250, 251].

It has been shown that T protects against hydrogen peroxide (H_2O_2)-induced apoptosis in the C2C12 muscle cell line [252]. Characteristical changes of apoptosis like nuclear fragmentation, cytoskeleton disorganization, and mitochondrial reorganization/dysfunction triggered by oxidative stress provoked by H_2O_2 are suppressed when cells are previously exposed to T. At short times of exposure to the apoptotic agent H_2O_2, muscle cells exhibit a defense response showing ERK2 phosphorylation, activation/phosphorylation of AKT and, consequently, the inactivation/phosphorylation of the proapoptotic protein BAD and an increase in HSP70 levels [91]. Thus, short-term stimulus with H_2O_2 induces the activation of the signaling pathways of cell mitogenic responses to promote cell survival. However, at longer times of treatment with the apoptotic agent, a decline in the phosphorylation of the proteins mentioned before, cytochrome c release and poly (ADP-ribose) polymerase (PARP) cleavage occur conducing finally to cell apoptosis. But if cells are treated with T prior to H_2O_2, a protective effect of the sex hormone is observed, which includes BAD inactivation (phosphorylation), a decrease in the proapoptotic protein Bax, and prevention of loss of mitochondrial membrane potential, pointing to the contribution of T in the modulation of the apoptotic intrinsic pathway and a protective effect of testosterone at mitochondrial level. Treatments with the AR antagonist, flutamide, prior to T and H_2O_2 exposure, decrease the protective effect of the sex steroid, suggesting the participation of the androgen receptor in the protective role of T. Although further studies are necessary to determine the molecular basis of sarcopenia related to states of T deficit, these data allow to beginning to elucidate the mechanism by which testosterone prevents apoptosis in skeletal muscle [91].

Cardiac Muscle

A physiological role of androgens in heart muscle is suggested by the presence of specific AR in the myocardium of rats, baboons, and rhesus monkeys. The affinity and binding capacity of myocardial AR are similar to those of skeletal muscle. However, androgen concentrations are higher in cardiac tissue than in striated muscle. Myocardial 5α-reductase activity is low, the same as skeletal muscle. Androgens can induce morphological and biochemical changes in heart muscle [1]. Electron microscopic studies of left ventricular tissue of rats treated with methandrostenolone revealed a significant increase of intermediate- sized non-myofibrillar filaments which may function as a cytoskeleton to provide support during ventricular contraction. Moreover, androgens potentiate GH-induced cardiac growth, whereas castration is associated with a reduced rate of aortic banding-induced cardiac hypertrophy. This effect was attributed to reduced nucleic acid and protein synthesis. Castration inhibits cardiac growth in rats and mice, and testosterone administration results in an increase in myocardial RNA and protein content, indicating an anabolic effect [1].

Substantial alterations in left ventricular function have been observed in rats using the isolated working rat heart preparation. Schaible *et al.* [253] found that castration resulted in an important downward shift in the mean force-velocity relationship at moderate and high left atrial pressures. The decrease in contractile function was related to a significant decrease in Ca^{++} myosin ATPase activity along with preferential accumulation of the V3 myosin isoenzyme and reduction of the VI isoenzyme. In mature female rats exercise-induced cardiac hypertrophy is not associated with any alterations in ventricular cytosol AR affinity or binding capacity, or serum concentrations of total or free T and DHT. Nevertheless, the relatively high androgen levels in female rats suggest that androgens may play an important role in exercise-induced cardiac hypertrophy [1].

Smooth Muscle

The smooth muscle of the guinea pig seminal vesicle is sensitive to androgens with castration resulting in a significant reduction in muscle weight and cell size, whereas cell number and collagen content do not change. These alterations could be reversed by 4 weeks of androgen treatment. Androgens also regulate the stimulatory effects of estrogen on seminal vesicle smooth muscles without changing estrogen receptor binding parameters [1]. The underlying mechanism of this cooperative interaction is not clear. The androgen sensitivity of smooth muscle tissue from other organs has not been studied extensively.

Skin

There are marked differences in the androgen concentrations of various skin sites. In both men and women, there are much higher concentrations of androgens in skin derived from the external genital anlage in the urogenital ridge, *e.g.* prepuce, clitoris, scrotum, and labia majora, than in nonsexual skin, *e.g.* thigh. In human skin, the androgen receptor expresses in epidermal keratinocytes, dermal papilla cells, hair follicles, sebaceous gland sebocytes, sweat glands, vascular endothelial and smooth muscle cells, and fibroblasts [254]. The skin contains all the enzymes needed to convert the adrenal androgens, dehydroepiandrosterone (DHEA) and dehydroepiandrosterone-sulfate (DHEAS), into DHT and therefore may be a local target of DHT-AR actions [254]. Besides, in androgen target skin, there is a high 5α-reductase activity, being DHT the principal androgen who exerts the AR mediated actions.

Studies in testicular feminization mice showed that they display reduced collagen content in the skin, kidney, and lung in comparison with male WT mice [255], pointing to the androgen/AR signaling involvement in the regulation of collagen synthesis in these tissues/organs.

Androgens are known to mediate sebum secretion and hair growth. Acne, androgenetic alopecia (male pattern baldness), and hirsutism are the three principal androgen-related diseases in dermatology [256, 257]. It has been shown that prepubertal castration prevents androgenetic alopecia and sebum secretion, while adult castration inhibits the progression of androgenic alopecia [256]. These castration effects can be reverted by T administration, suggesting that T is implicated in these dermatological diseases. Patients with no functional AR preserve the sebum profile of preadrenarchal children and never development acne, facial hair, or androgenetic alopecia [258].

AR Roles in the Development of Hair Follicles

In humans after puberty, androgens induce beard, axillary, and pubic hairs growth, but provoke vellus transformation in the frontal scalp hair in some genetic backgrounds, suggesting that the hair follicles have differential susceptibility to androgen/AR signaling. Contrary, hair follicles in the back of the head seemed to be insensitive to androgen/AR signaling [259].

Excess androgen production may be an etiological factor in hirsutism. Serum-free T concentration is increased in hirsute patients. The ovaries are the main source of serum androgens in these women. Although in most women with significant hirsutism, hyperandrogenemia can be documented, in some hirsute women serum total and free androgen concentrations are normal. It has been postulated that increased sensitivity of hair follicles to circulating androgens may explain such cases of idiopathic hirsutism, as supported by the demonstration of increased plasma levels and urinary excretion of 3α-androstanediol glucuronide in hirsute women [1], indicating increased target tissue metabolism of T *via* DHT. However, the possibility of enhanced nuclear translocation of androgen receptor cannot be excluded as a cause of increased androgen sensitivity.

In rat and mice, hair growth is induced by gonadectomy and inhibited by T [260, 261] and consequently they have androgen sensitivity similar to the frontal scalp of humans. ARKO mice grow longer and thicker hair than WT mice. DHT inhibited hair regrowth in WT, but not in ARKO mice, suggesting that AR is mediating the DHT effects [262]. Therefore, hair growth in response to DHT in mouse skin seems to resemble hair growth in the frontal scalp in men. However, in which cell type the AR is implicated and the molecular mechanisms of AR regulation remain unknown.

AR Roles in Cutaneous Wound Healing

Numerous studies have shown that androgen/AR signaling may play a significant role in cutaneous wound healing. Androgen deprivation by castration in mice results in accelerated wound healing related to attenuated inflammation that can be reverted by exogenous supplementation of androgen [263]. Furthermore, the 5α-reductase inhibitor accelerates wound healing similar to castration, dampening the local inflammatory response of the wounds [263]. These observations support the idea that androgen/AR signaling plays a suppressive role in cutaneous wound healing.

In male ARKO mice, cutaneous wound healing is accelerated with faster re-epithelialization, reduced inflammation, and augmented collagen deposition. Specific myeloid ARKO mice presented a similar acceleration of wound healing, pointing to a significant role of the AR in inflammatory cells, especially macrophages, in suppressing cutaneous wound repair [264]. TNFα, which is principally expressed in infiltrating macrophages, is decreased in ARKO wounds in comparison with WT wounds, whereas local restoration of TNFα could delay wound healing in ARKO mice, involving TNFα as a central mediator of AR function in wound healing suppression [265].

The role of the androgen receptor in keratinocytes and dermal fibroblasts in cutaneous wound healing were analyzed too. In Ker-ARKO mice, the wound healing rate was similar to WT, whereas, the re-epithelialization was delayed, suggesting that AR in keratinocytes inhibit the proliferation and/or migration of epithelial cells in the cutaneous wounds, but is not critical to suppress the overall repair process [265].

Sebaceous Glands

Androgens modulate the development and secretory activity of sebaceous glands on the face, upper back, and chest. Sebum production and histological size of facial sebaceous glands depend on T exposure and are suppressed by E2. As in skin, the major androgen in sebaceous glands is DHT, which is assumed to be the active hormone.

Positive correlations between acne and serum androgen levels have been reported by several groups. However, some women and men with severe acne do not have a high serum androgen concentration [1]. It is possible that the conversion *in situ* of dehydroepiandrosterone to a more potent androgen may explain some of the cases of acne in the absence of increased serum androgen levels.

Castration of male rats produces a striking reduction in the volume of the sebaceous glands, while subsequent administration of testosterone increases the volume of the glandular cells and enhanced cellular proliferation.

Sebum is composed of disintegrating sebaceous cells containing large amounts of lipids, and this lipid material is synthesized by the vesicular endoplasmic reticulum. Secchi *et al.* [266] have demonstrated that topical administration of antiandrogens markedly decreases the volume of the smooth endoplasmic reticulum in intact male rats and prevents the enlargement associated with testosterone treatment in castrated animals.

Summarizing, androgens have a direct effect on sebaceous gland growth and also modulate the quality of sebaceous secretions [1].

Adipose Tissue

Sex hormones seem to exert a role in the distribution of body fat, with males having a more proportion of upper body fat and females a greater proportion of lower body fat. Upper body fat appears to be identified with the metabolic complications of obesity. Evans *et al.* [267] have demonstrated that women with a preponderance of upper body fat have an increased free T compared to those with lower body fat preponderance (low waist to hips circumference ratio).

In castrated adults rats, testosterone replacement in low doses produces an increase in food intake and weight gain, whereas with high doses of the hormone there is a reduction in body weight. The decrease in body weight produced by testosterone appears to be due to its aromatization to estrogens. The major site of aromatization is adipose tissue. Both testosterone and estradiol reduce body weight by reducing fat mass. Both progesterone, which antagonizes the weight reducing actions of testosterone, and an aromatase inhibitor inhibit the weight-reducing effects of testosterone. Low doses of androgens increase adipose tissue lipoprotein lipase activity while high doses and estrogens inhibit the activity of this enzyme [1]. Reductions in lipoprotein lipase produced by high doses of testosterone are prevented by concurrent administration of an aromatase inhibitor. Thus, low doses of testosterone increase adipose tissue mass, and this effect does not involve 5α-reductase activity. In humans, this effect is seen predominantly in the upper torso and is associated with increased morbidity and mortality. High physiological doses of testosterone suppress adipose tissue mass and lipoprotein lipase as consequence of aromatization to estrogens.

With the employment of ARKO mice, it has been observed that the general loss of the AR leads to insulin and leptin resistance, dyslipidemia, and obesity. Lin *et al.* showed that, along with augmented body weight, male ARKO mice developed hyperinsulinemia, hyperleptinemia, hyperglycemia, hypoadiponectinemia, augmented serum levels of triglycerides and free fatty acids, and impaired glucose tolerance at advanced ages [268]. The recovery of the levels of androgen with DHT in ARKO male mice did not revert these changed parameters, suggesting that the observed phenotypes were a result of loss of AR function. The increment of lipid deposition in white adipose tissue was linked to the increase in the expression of 4 genes that stimulate late adipocyte differentiation and lipid accumulation in white adipose tissue of male ARKO mice in comparison with WT mice. These genes are the peroxisome proliferator-activated receptor gamma (*PPAR-γ*), CCAAT/enhancer-binding protein-α (*C/EBPα*), adipocyte fatty acid-binding protein 4/adipocyte protein 2 (*aP2*), and sterol regulatory element-binding protein 1c (*SREBP1c*). All together, these data suggest that loss of AR leads to insulin and leptin resistance, dysregulation of lipid metabolism and promote adipocyte differentiation and fat deposition resulting in obesity.

The late onset obesity and insulin resistance in ARKO mice seems to be in accordance to these phenotypes observed in humans with low androgen/AR signaling activity, such as hypogonadal men and PC patients treated with ADT. Furthermore, it has been presented that CAIS patients, principally those without orchidectomy, have higher propensity to develop obesity and abnormal insulin sensitivity than normal male and female subjects [269]. In the other hand, the lack of development of obesity and insulin resistance in female ARKO mice seems to suggest that metabolic responses to androgen/AR signaling are different between males and females.

An adipose tissue-specific ARKO model, reveals a phenotype of hyperleptinemia but no leptin resistance [270]. Of relevance, these mice are not obese, with normal body weight, adiposity index and adipocyte size. Another adipose tissue-specific ARKO model showed hyperinsulinemia in the absence of obesity [271]. The mice also had a greater weight response to a high fat diet. The global ARKO models suggest that fat mass is modulated by the androgen receptor. However, the lack of an obese phenotype in the adipose tissue-specific ARKO mice on a normal diet and the presence of a metabolic phenotype in muscle- and neuronal-specific ARKO models suggest that this modulation is happening through AR actions in other tissues than fat, including skeletal muscle and the brain/CNS.

CONSENT FOR PUBLICATION

Not applicable.

CONFLICT OF INTEREST

The authors confirm that this chapter contents have no conflict of interest.

ACKNOWLEDGEMENTS

National University of the South Argentina and National Research Council of Argentina (CONICET).

REFERENCES

[1]   Mooradian AD, Morley JE, Korenman SG. Biological actions of androgens. Endocr Rev 1987; 8(1): 1-28.
[http://dx.doi.org/10.1210/edrv-8-1-1] [PMID: 3549275]

[2]   Roy AK, Lavrovsky Y, Song CS, *et al.* Regulation of androgen action. Vitam Horm 1999; 55: 309-52.
[http://dx.doi.org/10.1016/S0083-6729(08)60938-3] [PMID: 9949684]

[3]   McLachlan RI, Wreford NG, O'Donnell L, de Kretser DM, Robertson DM. The endocrine regulation of spermatogenesis: independent roles for testosterone and FSH. J Endocrinol 1996; 148(1): 1-9.
[http://dx.doi.org/10.1677/joe.0.1480001] [PMID: 8568455]

[4]   Cunha GR, Donjacour AA, Cooke PS, *et al.* The endocrinology and developmental biology of the prostate. Endocr Rev 1987; 8(3): 338-62.
[http://dx.doi.org/10.1210/edrv-8-3-338] [PMID: 3308446]

[5]   Evans RM. The steroid and thyroid hormone receptor superfamily. Science 1988; 240(4854): 889-95.
[http://dx.doi.org/10.1126/science.3283939] [PMID: 3283939]

[6]   Laudet V. Evolution of the nuclear receptor superfamily: early diversification from an ancestral orphan receptor. J Mol Endocrinol 1997; 19(3): 207-26.
[http://dx.doi.org/10.1677/jme.0.0190207] [PMID: 9460643]

[7]   Eick GN, Thornton JW. Evolution of steroid receptors from an estrogen-sensitive ancestral receptor. Mol Cell Endocrinol 2011; 334(1-2): 31-8.
[http://dx.doi.org/10.1016/j.mce.2010.09.003] [PMID: 20837101]

[8]   Bain DL, Heneghan AF, Connaghan-Jones KD, Miura MT. Nuclear receptor structure: implications for function. Annu Rev Physiol 2007; 69: 201-20.
[http://dx.doi.org/10.1146/annurev.physiol.69.031905.160308] [PMID: 17137423]

[9]   Enmark E, Gustafsson JA. Orphan nuclear receptors the first eight years. Mol Endocrinol 1996; 10(11): 1293-307.
[PMID: 8923456]

[10]   Benoit G, Cooney A, Giguere V, *et al.* International Union of Pharmacology. LXVI. Orphan nuclear receptors. Pharmacol Rev 2006; 58(4): 798-836.
[http://dx.doi.org/10.1124/pr.58.4.10] [PMID: 17132856]

[11]   Mangelsdorf DJ, Thummel C, Beato M, *et al.* The nuclear receptor superfamily: the second decade. Cell 1995; 83(6): 835-9.
[http://dx.doi.org/10.1016/0092-8674(95)90199-X] [PMID: 8521507]

[12]   Kumar R, Betney R, Li J, Thompson EB, McEwan IJ. Induced alpha-helix structure in AF1 of the androgen receptor upon binding transcription factor TFIIF. Biochemistry 2004; 43(11): 3008-13.
[http://dx.doi.org/10.1021/bi035934p] [PMID: 15023052]

[13]   Yi P, Wu RC, Sandquist J, *et al.* Peptidyl-prolyl isomerase 1 (Pin1) serves as a coactivator of steroid receptor by regulating the activity of phosphorylated steroid receptor coactivator 3 (SRC-3/AIB1). Mol Cell Biol 2005; 25(21): 9687-99.
[http://dx.doi.org/10.1128/MCB.25.21.9687-9699.2005] [PMID: 16227615]

[14]   Migeon BR, Brown TR, Axelman J, Migeon CJ. Studies of the locus for androgen receptor: localization on the human X chromosome and evidence for homology with the Tfm locus in the mouse. Proc Natl Acad Sci USA 1981; 78(10): 6339-43.
[http://dx.doi.org/10.1073/pnas.78.10.6339] [PMID: 6947233]

[15]   Brown CJ, Goss SJ, Lubahn DB, *et al.* Androgen receptor locus on the human X chromosome: regional localization to Xq11-12 and description of a DNA polymorphism. Am J Hum Genet 1989; 44(2): 264-9.
[PMID: 2563196]

[16]   Gao W, Bohl CE, Dalton JT. Chemistry and structural biology of androgen receptor. Chem Rev 2005; 105(9): 3352-70.
[http://dx.doi.org/10.1021/cr020456u] [PMID: 16159155]

[17]   Beato M. Gene regulation by steroid hormones. Cell 1989; 56(3): 335-44.
[http://dx.doi.org/10.1016/0092-8674(89)90237-7] [PMID: 2644044]

[18]   Kuiper GG, Faber PW, van Rooij HC, *et al.* Structural organization of the human androgen receptor gene. J Mol Endocrinol 1989; 2(3): R1-4.
[http://dx.doi.org/10.1677/jme.0.002R001] [PMID: 2546571]

[19]   Lubahn DB, Brown TR, Simental JA, *et al.* Sequence of the intron/exon junctions of the coding region of the human androgen receptor gene and identification of a point mutation in a family with complete androgen insensitivity. Proc Natl Acad Sci USA 1989; 86(23): 9534-8.
[http://dx.doi.org/10.1073/pnas.86.23.9534] [PMID: 2594783]

[20]   Brann DW, Hendry LB, Mahesh VB. Emerging diversities in the mechanism of action of steroid hormones. J Steroid Biochem Mol Biol 1995; 52(2): 113-33.
[http://dx.doi.org/10.1016/0960-0760(94)00160-N] [PMID: 7873447]

[21]   Lieberherr M, Grosse B. Androgens increase intracellular calcium concentration and inositol 1,4,5-trisphosphate and diacylglycerol formation *via* a pertussis toxin-sensitive G-protein. J Biol Chem 1994; 269(10): 7217-23.
[PMID: 8125934]

[22]   Fischer L, Catz D, Kelley D. An androgen receptor mRNA isoform associated with hormone-induced cell proliferation. Proc Natl Acad Sci USA 1993; 90(17): 8254-8.
[http://dx.doi.org/10.1073/pnas.90.17.8254] [PMID: 7690145]

[23]   Dehm SM, Tindall DJ. Androgen receptor structural and functional elements: role and regulation in prostate cancer. Mol Endocrinol 2007; 21(12): 2855-63.
[http://dx.doi.org/10.1210/me.2007-0223] [PMID: 17636035]

[24]   Hiipakka RA, Liao S. Molecular mechanism of androgen action. Trends Endocrinol Metab 1998; 9(8): 317-24.
[http://dx.doi.org/10.1016/S1043-2760(98)00081-2] [PMID: 18406296]

[25]   Fragkaki AG, Angelis YS, Koupparis M, Tsantili-Kakoulidou A, Kokotos G, Georgakopoulos C. Structural characteristics of anabolic androgenic steroids contributing to binding to the androgen receptor and to their anabolic and androgenic activities. Applied modifications in the steroidal structure. Steroids 2009; 74(2): 172-97.
[http://dx.doi.org/10.1016/j.steroids.2008.10.016] [PMID: 19028512]

[26]   Heinlein CA, Chang C. Role of chaperones in nuclear translocation and transactivation of steroid receptors. Endocrine 2001; 14(2): 143-9.
[http://dx.doi.org/10.1385/ENDO:14:2:143] [PMID: 11394630]

[27]   Fang Y, Fliss AE, Robins DM, Caplan AJ. Hsp90 regulates androgen receptor hormone binding affinity *in vivo*. J Biol Chem 1996; 271(45): 28697-702.
[http://dx.doi.org/10.1074/jbc.271.45.28697] [PMID: 8910505]

[28]   Georget V, Térouanne B, Nicolas JC, Sultan C. Mechanism of antiandrogen action: key role of hsp90 in conformational change and transcriptional activity of the androgen receptor. Biochemistry 2002; 41(39): 11824-31.
[http://dx.doi.org/10.1021/bi0259150] [PMID: 12269826]

[29]   Jenster G, Spencer TE, Burcin MM, Tsai SY, Tsai M-J, O'Malley BW. Steroid receptor induction of gene transcription: a two-step model. Proc Natl Acad Sci USA 1997; 94(15): 7879-84.
[http://dx.doi.org/10.1073/pnas.94.15.7879] [PMID: 9223281]

[30]   Ikonen T, Palvimo JJ, Jänne OA. Interaction between the amino- and carboxyl-terminal regions of the rat androgen receptor modulates transcriptional activity and is influenced by nuclear receptor

coactivators. J Biol Chem 1997; 272(47): 29821-8.
[http://dx.doi.org/10.1074/jbc.272.47.29821] [PMID: 9368054]

[31]   Heery DM, Kalkhoven E, Hoare S, Parker MG. A signature motif in transcriptional co-activators mediates binding to nuclear receptors. Nature 1997; 387(6634): 733-6.
[http://dx.doi.org/10.1038/42750] [PMID: 9192902]

[32]   Shiau AK, Barstad D, Loria PM, *et al*. The structural basis of estrogen receptor/coactivator recognition and the antagonism of this interaction by tamoxifen. Cell 1998; 95(7): 927-37.
[http://dx.doi.org/10.1016/S0092-8674(00)81717-1] [PMID: 9875847]

[33]   Jenster G, Trapman J, Brinkmann AO. Nuclear import of the human androgen receptor. Biochem J 1993; 293(Pt 3): 761-8.
[http://dx.doi.org/10.1042/bj2930761] [PMID: 8352744]

[34]   Zhou Z-X, Sar M, Simental JA, Lane MV, Wilson EM. A ligand-dependent bipartite nuclear targeting signal in the human androgen receptor. Requirement for the DNA-binding domain and modulation by NH2-terminal and carboxyl-terminal sequences. J Biol Chem 1994; 269(18): 13115-23.
[PMID: 8175737]

[35]   van Steensel B, Jenster G, Damm K, Brinkmann AO, van Driel R. Domains of the human androgen receptor and glucocorticoid receptor involved in binding to the nuclear matrix. J Cell Biochem 1995; 57(3): 465-78.
[http://dx.doi.org/10.1002/jcb.240570312] [PMID: 7768981]

[36]   Young CY, Montgomery BT, Andrews PE, Qui SD, Bilhartz DL, Tindall DJ. Hormonal regulation of prostate-specific antigen messenger RNA in human prostatic adenocarcinoma cell line LNCaP. Cancer Res 1991; 51(14): 3748-52.
[PMID: 1712248]

[37]   Guiochon-Mantel A, Delabre K, Lescop P, Milgrom E. Nuclear localization signals also mediate the outward movement of proteins from the nucleus. Proc Natl Acad Sci USA 1994; 91(15): 7179-83.
[http://dx.doi.org/10.1073/pnas.91.15.7179] [PMID: 8041765]

[38]   Tyagi RK, Amazit L, Lescop P, Milgrom E, Guiochon-Mantel A. Mechanisms of progesterone receptor export from nuclei: role of nuclear localization signal, nuclear export signal, and ran guanosine triphosphate. Mol Endocrinol 1998; 12(11): 1684-95.
[http://dx.doi.org/10.1210/mend.12.11.0197] [PMID: 9817595]

[39]   Mangelsdorf DJ, Thummel C, Beato M, *et al*. The nuclear receptor superfamily: the second decade. Cell 1995; 83(6): 835-9.
[http://dx.doi.org/10.1016/0092-8674(95)90199-X] [PMID: 8521507]

[40]   McKenna NJ, Lanz RB, O'Malley BW. Nuclear receptor coregulators: cellular and molecular biology. Endocr Rev 1999; 20(3): 321-44.
[PMID: 10368774]

[41]   Bourguet W, Germain P, Gronemeyer H. Nuclear receptor ligand-binding domains: three-dimensional structures, molecular interactions and pharmacological implications. Trends Pharmacol Sci 2000; 21(10): 381-8.
[http://dx.doi.org/10.1016/S0165-6147(00)01548-0] [PMID: 11050318]

[42]   Nelson CC, Hendy SC, Shukin RJ, *et al*. Determinants of DNA sequence specificity of the androgen, progesterone, and glucocorticoid receptors: evidence for differential steroid receptor response elements. Mol Endocrinol 1999; 13(12): 2090-107.
[http://dx.doi.org/10.1210/mend.13.12.0396] [PMID: 10598584]

[43]   Kallio PJ, Jänne OA, Palvimo JJ. Agonists, but not antagonists, alter the conformation of the hormone-binding domain of androgen receptor. Endocrinology 1994; 134(2): 998-1001.
[http://dx.doi.org/10.1210/endo.134.2.8299593] [PMID: 8299593]

[44]   Palvimo JJ, Reinikainen P, Ikonen T, Kallio PJ, Moilanen A, Jänne OA. Mutual transcriptional

interference between RelA and androgen receptor. J Biol Chem 1996; 271(39): 24151-6.
[http://dx.doi.org/10.1074/jbc.271.39.24151] [PMID: 8798655]

[45]   Adler AJ, Danielsen M, Robins DM. Androgen-specific gene activation *via* a consensus glucocorticoid response element is determined by interaction with nonreceptor factors. Proc Natl Acad Sci USA 1992; 89(24): 11660-3.
[http://dx.doi.org/10.1073/pnas.89.24.11660] [PMID: 1465381]

[46]   Schoenmakers E, Verrijdt G, Peeters B, Verhoeven G, Rombauts W, Claessens F. Differences in DNA binding characteristics of the androgen and glucocorticoid receptors can determine hormone-specific responses. J Biol Chem 2000; 275(16): 12290-7.
[http://dx.doi.org/10.1074/jbc.275.16.12290] [PMID: 10766868]

[47]   Foradori CD, Weiser MJ, Handa RJ. Non-genomic actions of androgens. Front Neuroendocrinol 2008; 29(2): 169-81.
[http://dx.doi.org/10.1016/j.yfrne.2007.10.005] [PMID: 18093638]

[48]   Lösel R, Wehling M. Nongenomic actions of steroid hormones. Nat Rev Mol Cell Biol 2003; 4(1): 46-56.
[http://dx.doi.org/10.1038/nrm1009] [PMID: 12511868]

[49]   Simoncini T, Genazzani AR. Non-genomic actions of sex steroid hormones. Eur J Endocrinol 2003; 148(3): 281-92.
[http://dx.doi.org/10.1530/eje.0.1480281] [PMID: 12611608]

[50]   Levin ER. Integration of the extranuclear and nuclear actions of estrogen. Mol Endocrinol 2005; 19(8): 1951-9.
[http://dx.doi.org/10.1210/me.2004-0390] [PMID: 15705661]

[51]   Migliaccio A, Castoria G, Di Domenico M, *et al.* Steroid-induced androgen receptor-oestradiol receptor beta-Src complex triggers prostate cancer cell proliferation. EMBO J 2000; 19(20): 5406-17.
[http://dx.doi.org/10.1093/emboj/19.20.5406] [PMID: 11032808]

[52]   Silva FR, Leite LD, Wassermann GF. Rapid signal transduction in Sertoli cells. Eur J Endocrinol 2002; 147(3): 425-33.
[http://dx.doi.org/10.1530/eje.0.1470425] [PMID: 12213681]

[53]   Cato AC, Nestl A, Mink S. Rapid actions of steroid receptors in cellular signaling pathways. Sci STKE 2002; 2002(138): re9.
[PMID: 12084906]

[54]   Fix C, Jordan C, Cano P, Walker WH. Testosterone activates mitogen-activated protein kinase and the cAMP response element binding protein transcription factor in Sertoli cells. Proc Natl Acad Sci USA 2004; 101(30): 10919-24.
[http://dx.doi.org/10.1073/pnas.0404278101] [PMID: 15263086]

[55]   Kang HY, Cho CL, Huang KL, *et al.* Nongenomic androgen activation of phosphatidylinositol 3-kinase/Akt signaling pathway in MC3T3-E1 osteoblasts. J Bone Miner Res 2004; 19(7): 1181-90.
[http://dx.doi.org/10.1359/JBMR.040306] [PMID: 15177002]

[56]   Cheng J, Watkins SC, Walker WH. Testosterone activates mitogen-activated protein kinase *via* Src kinase and the epidermal growth factor receptor in sertoli cells. Endocrinology 2007; 148(5): 2066-74.
[http://dx.doi.org/10.1210/en.2006-1465] [PMID: 17272394]

[57]   Cinar B, Mukhopadhyay NK, Meng G, Freeman MR. Phosphoinositide 3-kinase-independent non-genomic signals transit from the androgen receptor to Akt1 in membrane raft microdomains. J Biol Chem 2007; 282(40): 29584-93.
[http://dx.doi.org/10.1074/jbc.M703310200] [PMID: 17635910]

[58]   Agoulnik IU, Bingman WE III, Nakka M, *et al.* Target gene-specific regulation of androgen receptor activity by p42/p44 mitogen-activated protein kinase. Mol Endocrinol 2008; 22(11): 2420-32.
[http://dx.doi.org/10.1210/me.2007-0481] [PMID: 18787043]

[59] Yu J, Akishita M, Eto M, *et al.* Src kinase-mediates androgen receptor-dependent non-genomic activation of signaling cascade leading to endothelial nitric oxide synthase. Biochem Biophys Res Commun 2012; 424(3): 538-43.
[http://dx.doi.org/10.1016/j.bbrc.2012.06.151] [PMID: 22771325]

[60] Rahman F, Christian HC. Non-classical actions of testosterone: an update. Trends Endocrinol Metab 2007; 18(10): 371-8.
[http://dx.doi.org/10.1016/j.tem.2007.09.004] [PMID: 17997105]

[61] Benten WP, Becker A, Schmitt-Wrede HP, Wunderlich F. Developmental regulation of intracellular and surface androgen receptors in T cells. Steroids 2002; 67(11): 925-31.
[http://dx.doi.org/10.1016/S0039-128X(02)00055-7] [PMID: 12234628]

[62] Sun M, Yang L, Feldman RI, *et al.* Activation of phosphatidylinositol 3-kinase/Akt pathway by androgen through interaction of p85alpha, androgen receptor, and Src. J Biol Chem 2003; 278(44): 42992-3000.
[http://dx.doi.org/10.1074/jbc.M306295200] [PMID: 12933816]

[63] Heinlein CA, Chang C. The roles of androgen receptors and androgen-binding proteins in nongenomic androgen actions. Mol Endocrinol 2002; 16(10): 2181-7.
[http://dx.doi.org/10.1210/me.2002-0070] [PMID: 12351684]

[64] Kousteni S, Bellido T, Plotkin LI, *et al.* Nongenotropic, sex-nonspecific signaling through the estrogen or androgen receptors: dissociation from transcriptional activity. Cell 2001; 104(5): 719-30.
[PMID: 11257226]

[65] Boonyaratanakornkit V, Scott MP, Ribon V, *et al.* Progesterone receptor contains a proline-rich motif that activates c-Src family tyrosine kinases. Mol Cell 2001; 8: 269-80.
[http://dx.doi.org/10.1016/S1097-2765(01)00304-5] [PMID: 11545730]

[66] Konoplya EF, Popoff EH. Identification of the classical androgen receptor in male rat liver and prostate cell plasma membranes. Int J Biochem 1992; 24(12): 1979-83.
[http://dx.doi.org/10.1016/0020-711X(92)90294-B] [PMID: 1473610]

[67] Benten WPM, Lieberherr M, Giese G, *et al.* Functional testosterone receptors in plasma membranes of T cells. FASEB J 1999; 13(1): 123-33.
[http://dx.doi.org/10.1096/fasebj.13.1.123] [PMID: 9872937]

[68] Kampa M, Papakonstanti EA, Hatzoglou A, Stathopoulos EN, Stournaras C, Castanas E. The human prostate cancer cell line LNCaP bears functional membrane testosterone receptors that increase PSA secretion and modify actin cytoskeleton. FASEB J 2002; 16(11): 1429-31.
[http://dx.doi.org/10.1096/fj.02-0131fje] [PMID: 12205037]

[69] Estrada M, Espinosa A, Müller M, Jaimovich E. Testosterone stimulates intracellular calcium release and mitogen-activated protein kinases *via* a G protein-coupled receptor in skeletal muscle cells. Endocrinology 2003; 144(8): 3586-97.
[http://dx.doi.org/10.1210/en.2002-0164] [PMID: 12865341]

[70] Gill A, Jamnongjit M, Hammes SR. Androgens promote maturation and signaling in mouse oocytes independent of transcription: a release of inhibition model for mammalian oocyte meiosis. Mol Endocrinol 2004; 18(1): 97-104.
[http://dx.doi.org/10.1210/me.2003-0326] [PMID: 14576339]

[71] Papadopoulou N, Charalampopoulos I, Anagnostopoulou V, *et al.* Membrane androgen receptor activation triggers down-regulation of PI-3K/Akt/NF-kappaB activity and induces apoptotic responses *via* Bad, FasL and caspase-3 in DU145 prostate cancer cells. Mol Cancer 2008; 7: 88.
[http://dx.doi.org/10.1186/1476-4598-7-88] [PMID: 19055752]

[72] Takeda H, Chodak G, Mutchnik S, Nakamoto T, Chang C. Immunohistochemical localization of androgen receptors with mono- and polyclonal antibodies to androgen receptor. J Endocrinol 1990; 126(1): 17-25.

[http://dx.doi.org/10.1677/joe.0.1260017] [PMID: 2199591]

[73]   Mainwaring WIP, Mangan FR. A study of the androgen receptors in a variety of androgen-sensitive tissues. J Endocrinol 1973; 59(1): 121-39.
[http://dx.doi.org/10.1677/joe.0.0590121] [PMID: 4356005]

[74]   Mayer M, Rosen F. Effect of endocrine manipulations on glucocorticoid binding capacity in rat skeletal muscle. Acta Endocrinol (Copenh) 1978; 88(1): 199-208.
[http://dx.doi.org/10.1530/acta.0.0880199] [PMID: 148198]

[75]   Snochowski M, Dahlberg E, Gustafsson J-A. Characterization and quantification of the androgen and glucocorticoid receptors in cytosol from rat skeletal muscle. Eur J Biochem 1980; 111(2): 603-16.
[http://dx.doi.org/10.1111/j.1432-1033.1980.tb04977.x] [PMID: 6970125]

[76]   Dahlberg E, Snochowski M, Gustafsson J-A. Regulation of the androgen and glucocorticoid receptors in rat and mouse skeletal muscle cytosol. Endocrinology 1981; 108(4): 1431-40.
[http://dx.doi.org/10.1210/endo-108-4-1431] [PMID: 6970661]

[77]   Max SR. Cytosolic androgen receptor in skeletal muscle from normal and testicular feminization mutant (Tfm) rats. Biochem Biophys Res Commun 1981; 101(3): 792-9.
[http://dx.doi.org/10.1016/0006-291X(81)91820-9] [PMID: 6975622]

[78]   Max SR, Mufti S, Carlson BM. Cytosolic androgen receptor in regenerating rat levator ani muscle. Biochem J 1981; 200(1): 77-82.
[http://dx.doi.org/10.1042/bj2000077] [PMID: 6977357]

[79]   Michel G, Baulieu E-E. Androgen receptor in rat skeletal muscle: characterization and physiological variations. Endocrinology 1980; 107(6): 2088-98.
[http://dx.doi.org/10.1210/endo-107-6-2088] [PMID: 6968675]

[80]   Estrada M, Espinosa A, Müller M, Jaimovich E. Testosterone stimulates intracellular calcium release and mitogen-activated protein kinases *via* a G protein-coupled receptor in skeletal muscle cells. Endocrinology 2003; 144(8): 3586-97.
[http://dx.doi.org/10.1210/en.2002-0164] [PMID: 12865341]

[81]   Monje P, Boland R. Subcellular distribution of native estrogen receptor alpha and beta isoforms in rabbit uterus and ovary. J Cell Biochem 2001; 82(3): 467-79.
[http://dx.doi.org/10.1002/jcb.1182] [PMID: 11500923]

[82]   Cammarata PR, Chu S, Moor A, Wang Z, Yang SH, Simpkins JW. Subcellular distribution of native estrogen receptor alpha and beta subtypes in cultured human lens epithelial cells. Exp Eye Res 2004; 78(4): 861-71.
[http://dx.doi.org/10.1016/j.exer.2003.09.027] [PMID: 15037120]

[83]   Chen JQ, Delannoy M, Cooke C, Yager JD. Mitochondrial localization of ERalpha and ERbeta in human MCF7 cells. Am J Physiol Endocrinol Metab 2004; 286(6): E1011-22.
[http://dx.doi.org/10.1152/ajpendo.00508.2003] [PMID: 14736707]

[84]   Yang SH, Liu R, Perez EJ, *et al.* Mitochondrial localization of estrogen receptor beta. Proc Natl Acad Sci USA 2004; 101(12): 4130-5.
[http://dx.doi.org/10.1073/pnas.0306948101] [PMID: 15024130]

[85]   Milanesi L, Russo de Boland A, Boland R. Expression and localization of estrogen receptor alpha in the C2C12 murine skeletal muscle cell line. J Cell Biochem 2008; 104(4): 1254-73.
[http://dx.doi.org/10.1002/jcb.21706] [PMID: 18348185]

[86]   Solakidi S, Psarra AM, Nikolaropoulos S, Sekeris CE. Estrogen receptors alpha and beta (ERalpha and ERbeta) and androgen receptor (AR) in human sperm: localization of ERbeta and AR in mitochondria of the midpiece. Hum Reprod 2005; 20(12): 3481-7.
[http://dx.doi.org/10.1093/humrep/dei267] [PMID: 16123086]

[87]   Yaffe D, Saxel O. Serial passaging and differentiation of myogenic cells isolated from dystrophic mouse muscle. Nature 1977; 270(5639): 725-7.

[http://dx.doi.org/10.1038/270725a0] [PMID: 563524]

[88] Yoshida N, Yoshida S, Koishi K, Masuda K, Nabeshima Y. Cell heterogeneity upon myogenic differentiation: down-regulation of MyoD and Myf-5 generates 'reserve cells'. J Cell Sci 1998; 111(Pt 6): 769-79.
[PMID: 9472005]

[89] Pronsato L, Boland R, Milanesi L. Testosterone exerts antiapoptotic effects against H_2O_2 in C2C12 skeletal muscle cells through the apoptotic intrinsic pathway. J Endocrinol 2012; 212(3): 371-81.
[http://dx.doi.org/10.1530/JOE-11-0234] [PMID: 22219300]

[90] Pronsato L, Boland R, Milanesi L. Non-classical localization of androgen receptor in the C2C12 skeletal muscle cell line. Arch Biochem Biophys 2013; 530(1): 13-22.
[http://dx.doi.org/10.1016/j.abb.2012.12.011] [PMID: 23262317]

[91] Guo Z, Qiu Y. A new trick of an old molecule: androgen receptor splice variants taking the stage?! Int J Biol Sci 2011; 7(6): 815-22.
[http://dx.doi.org/10.7150/ijbs.7.815] [PMID: 21750650]

[92] Benten WP, Lieberherr M, Stamm O, Wrehlke C, Guo Z, Wunderlich F. Testosterone signaling through internalizable surface receptors in androgen receptor-free macrophages. Mol Biol Cell 1999; 10(10): 3113-23.
[http://dx.doi.org/10.1091/mbc.10.10.3113] [PMID: 10512854]

[93] Wang Z, Liu L, Hou J, *et al.* Rapid membrane effect of testosterone in LNCaP cells. Urol Int 2008; 81(3): 353-9.
[http://dx.doi.org/10.1159/000151418] [PMID: 18931557]

[94] Kampa M, Kogia C, Theodoropoulos PA, *et al.* Activation of membrane androgen receptors potentiates the antiproliferative effects of paclitaxel on human prostate cancer cells. Mol Cancer Ther 2006; 5(5): 1342-51.
[http://dx.doi.org/10.1158/1535-7163.MCT-05-0527] [PMID: 16731768]

[95] Hatzoglou A, Kampa M, Kogia C, *et al.* Membrane androgen receptor activation induces apoptotic regression of human prostate cancer cells *in vitro* and *in vivo*. J Clin Endocrinol Metab 2005; 90(2): 893-903.
[http://dx.doi.org/10.1210/jc.2004-0801] [PMID: 15585562]

[96] Gu S, Papadopoulou N, Gehring EM, *et al.* Functional membrane androgen receptors in colon tumors trigger pro-apoptotic responses *in vitro* and reduce drastically tumor incidence *in vivo*. Mol Cancer 2009; 8: 114.
[http://dx.doi.org/10.1186/1476-4598-8-114] [PMID: 19948074]

[97] Papadopoulou N, Papakonstanti EA, Kallergi G, Alevizopoulos K, Stournaras C. Membrane androgen receptor activation in prostate and breast tumor cells: molecular signaling and clinical impact. IUBMB Life 2009; 61(1): 56-61.
[http://dx.doi.org/10.1002/iub.150] [PMID: 19109827]

[98] Gatson JW, Kaur P, Singh M. Dihydrotestosterone differentially modulates the mitogen-activated protein kinase and the phosphoinositide 3-kinase/Akt pathways through the nuclear and novel membrane androgen receptor in C6 cells. Endocrinology 2006; 147(4): 2028-34.
[http://dx.doi.org/10.1210/en.2005-1395] [PMID: 16410299]

[99] Alexaki VI, Charalampopoulos I, Kampa M, *et al.* Activation of membrane estrogen receptors induce pro-survival kinases. J Steroid Biochem Mol Biol 2006; 98(2-3): 97-110.
[http://dx.doi.org/10.1016/j.jsbmb.2005.08.017] [PMID: 16414261]

[100] Somjen D, Kohen F, Gayer B, Kulik T, Knoll E, Stern N. Role of putative membrane receptors in the effect of androgens on human vascular cell growth. J Endocrinol 2004; 180(1): 97-106.
[http://dx.doi.org/10.1677/joe.0.1800097] [PMID: 14709148]

[101] Ruizeveld de Winter JA, Trapman J, Vermey M, Mulder E, Zegers ND, van der Kwast TH. Androgen

receptor expression in human tissues: an immunohistochemical study. J Histochem Cytochem 1991; 39(7): 927-36.
[http://dx.doi.org/10.1177/39.7.1865110] [PMID: 1865110]

[102] Iwamura M, Abrahamsson PA, Benning CM, Cockett AT, di Sant'Agnese PA. Androgen receptor immunostaining and its tissue distribution in formalin-fixed, paraffin-embedded sections after microwave treatment. J Histochem Cytochem 1994; 42(6): 783-8.
[http://dx.doi.org/10.1177/42.6.8189040] [PMID: 8189040]

[103] Griffin JE. Androgen resistance--the clinical and molecular spectrum. N Engl J Med 1992; 326(9): 611-8.
[http://dx.doi.org/10.1056/NEJM199202273260906] [PMID: 1734252]

[104] Kyprianou N, Isaacs JT. Activation of programmed cell death in the rat ventral prostate after castration. Endocrinology 1988; 122(2): 552-62.
[http://dx.doi.org/10.1210/endo-122-2-552] [PMID: 2828003]

[105] Wu CP, Gu FL. The prostate in eunuchs. Prog Clin Biol Res 1991; 370: 249-55.
[PMID: 1924456]

[106] Pollard M, Luckert PH, Sporn MB. Prevention of primary prostate cancer in Lobund-Wistar rats by N-(4-hydroxyphenyl)retinamide. Cancer Res 1991; 51(13): 3610-1.
[PMID: 1829024]

[107] Pollard M, Luckert PH. The inhibitory effect of 4-hydroxyphenyl retinamide (4-HPR) on metastasis of prostate adenocarcinoma-III cells in Lobund-Wistar rats. Cancer Lett 1991; 59(2): 159-63.
[http://dx.doi.org/10.1016/0304-3835(91)90181-G] [PMID: 1832081]

[108] Stanbrough M, Leav I, Kwan PW, Bubley GJ, Balk SP. Prostatic intraepithelial neoplasia in mice expressing an androgen receptor transgene in prostate epithelium. Proc Natl Acad Sci USA 2001; 98(19): 10823-8.
[http://dx.doi.org/10.1073/pnas.191235898] [PMID: 11535819]

[109] Yeh S, Tsai MY, Xu Q, *et al.* Generation and characterization of androgen receptor knockout (ARKO) mice: an *in vivo* model for the study of androgen functions in selective tissues. Proc Natl Acad Sci USA 2002; 99(21): 13498-503.
[http://dx.doi.org/10.1073/pnas.212474399] [PMID: 12370412]

[110] Notini AJ, Davey RA, McManus JF, Bate KL, Zajac JD. Genomic actions of the androgen receptor are required for normal male sexual differentiation in a mouse *model.* J Mol Endocrinol 2005; 35(3): 547-55.
[http://dx.doi.org/10.1677/jme.1.01884] [PMID: 16326839]

[111] Simanainen U, Allan CM, Lim P, *et al.* Disruption of prostate epithelial androgen receptor impedes prostate lobe-specific growth and function. Endocrinology 2007; 148(5): 2264-72.
[http://dx.doi.org/10.1210/en.2006-1223] [PMID: 17317769]

[112] Simanainen U, McNamara K, Gao YR, *et al.* Anterior prostate epithelial AR inactivation modifies estrogen receptor expression and increases estrogen sensitivity. Am J Physiol Endocrinol Metab 2011; 301(4): E727-35.
[http://dx.doi.org/10.1152/ajpendo.00580.2010] [PMID: 21750267]

[113] Wu CT, Altuwaijri S, Ricke WA, *et al.* Increased prostate cell proliferation and loss of cell differentiation in mice lacking prostate epithelial androgen receptor. Proc Natl Acad Sci USA 2007; 104(31): 12679-84.
[http://dx.doi.org/10.1073/pnas.0704940104] [PMID: 17652515]

[114] Welsh M, Moffat L, McNeilly A, *et al.* Smooth muscle cell-specific knockout of androgen receptor: a new model for prostatic disease. Endocrinology 2011; 152(9): 3541-51.
[http://dx.doi.org/10.1210/en.2011-0282] [PMID: 21733831]

[115] Yu S, Zhang C, Lin CC, *et al.* Altered prostate epithelial development and IGF-1 signal in mice

lacking the androgen receptor in stromal smooth muscle cells. Prostate 2011; 71(5): 517-24.
[http://dx.doi.org/10.1002/pros.21264] [PMID: 20945497]

[116]   Lai KP, Yamashita S, Vitkus S, Shyr CR, Yeh S, Chang C. Suppressed prostate epithelial development with impaired branching morphogenesis in mice lacking stromal fibromuscular androgen receptor. Mol Endocrinol 2012; 26(1): 52-66.
[http://dx.doi.org/10.1210/me.2011-1189] [PMID: 22135068]

[117]   Huggins C, Hodges CV. Studies on prostatic cancer: I The effects of castration, of estrogen and of androgen injection on serum phosphatases in metastatic carcinoma of the prostate. Cancer Res 1941; 1: 293-7.

[118]   Grossmann M, Cheung AS, Zajac JD. Androgens and prostate cancer; pathogenesis and deprivation therapy. Best Pract Res Clin Endocrinol Metab 2013; 27(4): 603-16.
[http://dx.doi.org/10.1016/j.beem.2013.05.001] [PMID: 24054933]

[119]   Beltran H, Beer TM, Carducci MA, *et al.* New therapies for castration-resistant prostate cancer: efficacy and safety. Eur Urol 2011; 60(2): 279-90.
[http://dx.doi.org/10.1016/j.eururo.2011.04.038] [PMID: 21592649]

[120]   Knudsen KE, Scher HI. Starving the addiction: new opportunities for durable suppression of AR signaling in prostate cancer. Clin Cancer Res 2009; 15(15): 4792-8.
[http://dx.doi.org/10.1158/1078-0432.CCR-08-2660] [PMID: 19638458]

[121]   Palmberg C, Koivisto P, Hyytinen E, *et al.* Androgen receptor gene amplification in a recurrent prostate cancer after monotherapy with the nonsteroidal potent antiandrogen Casodex (bicalutamide) with a subsequent favorable response to maximal androgen blockade. Eur Urol 1997; 31(2): 216-9.
[http://dx.doi.org/10.1159/000474453] [PMID: 9076469]

[122]   Marcelli M, Ittmann M, Mariani S, *et al.* Androgen receptor mutations in prostate cancer. Cancer Res 60:944-949, 2000 167. Barrack ER: Androgen receptor mutations in prostate cancer. Mt Sinai J Med 1996; 63: 403-12.
[PMID: 8898547]

[123]   Culig Z, Hobisch A, Hittmair A, *et al.* Androgen receptor gene mutations in prostate cancer. Implications for disease progression and therapy. Drugs Aging 1997; 10(1): 50-8.
[http://dx.doi.org/10.2165/00002512-199710010-00005] [PMID: 9111707]

[124]   Suzuki H, Sato N, Watabe Y, Masai M, Seino S, Shimazaki J. Androgen receptor gene mutations in human prostate cancer. J Steroid Biochem Mol Biol 1993; 46(6): 759-65.
[http://dx.doi.org/10.1016/0960-0760(93)90316-O] [PMID: 8274409]

[125]   Wallén MJ, Linja M, Kaartinen K, Schleutker J, Visakorpi T. Androgen receptor gene mutations in hormone-refractory prostate cancer. J Pathol 1999; 189(4): 559-63.
[http://dx.doi.org/10.1002/(SICI)1096-9896(199912)189:4<559::AID-PATH471>3.0.CO;2-Y]
[PMID: 10629558]

[126]   Taplin ME, Bubley GJ, Shuster TD, *et al.* Mutation of the androgen-receptor gene in metastatic androgen-independent prostate cancer. N Engl J Med 1995; 332(21): 1393-8.
[http://dx.doi.org/10.1056/NEJM199505253322101] [PMID: 7723794]

[127]   Tan J, Sharief Y, Hamil KG, *et al.* Dehydroepiandrosterone activates mutant androgen receptors expressed in the androgen-dependent human prostate cancer xenograft CWR22 and LNCaP cells. Mol Endocrinol 1997; 11(4): 450-9.
[http://dx.doi.org/10.1210/mend.11.4.9906] [PMID: 9092797]

[128]   Matias PM, Donner P, Coelho R, *et al.* Structural evidence for ligand specificity in the binding domain of the human androgen receptor. Implications for pathogenic gene mutations. J Biol Chem 2000; 275(34): 26164-71.
[http://dx.doi.org/10.1074/jbc.M004571200] [PMID: 10840043]

[129]   Gregory CW, Johnson RT Jr, Mohler JL, French FS, Wilson EM. Androgen receptor stabilization in

recurrent prostate cancer is associated with hypersensitivity to low androgen. Cancer Res 2001; 61(7): 2892-8.
[PMID: 11306464]

[130] Gregory CW, He B, Johnson RT, *et al.* A mechanism for androgen receptor-mediated prostate cancer recurrence after androgen deprivation therapy. Cancer Res 2001; 61(11): 4315-9.
[PMID: 11389051]

[131] Buchanan G, Yang M, Harris JM, *et al.* Mutations at the boundary of the hinge and ligand binding domain of the androgen receptor confer increased transactivation function. Mol Endocrinol 2001; 15(1): 46-56.
[http://dx.doi.org/10.1210/mend.15.1.0581] [PMID: 11145738]

[132] Wang Q, Lu J, Yong EL. Ligand- and coactivator-mediated transactivation function (AF2) of the androgen receptor ligand-binding domain is inhibited by the cognate hinge region. J Biol Chem 2001; 276(10): 7493-9.
[http://dx.doi.org/10.1074/jbc.M009916200] [PMID: 11102454]

[133] McEwen BS. Binding and metabolism of sex steroids by the hypothalamic-pituitary unit: physiological implications. Annu Rev Physiol 1980; 42: 97-110.
[http://dx.doi.org/10.1146/annurev.ph.42.030180.000525] [PMID: 6996603]

[134] Raisman G, Field PM. Sexual dimorphism in the neuropil of the preoptic area of the rat and its dependence on neonatal androgen. Brain Res 1973; 54: 1-29.
[http://dx.doi.org/10.1016/0006-8993(73)90030-9] [PMID: 4122682]

[135] Toran-Allerand CD. Sex steroids and the development of the newborn mouse hypothalamus and preoptic area *in vitro*: implications for sexual differentiation. Brain Res 1976; 106(2): 407-12.
[http://dx.doi.org/10.1016/0006-8993(76)91038-6] [PMID: 1276881]

[136] Nottebohm F, Arnold AP. Sexual dimorphism in vocal control areas of the songbird brain. Science 1976; 194(4261): 211-3.
[http://dx.doi.org/10.1126/science.959852] [PMID: 959852]

[137] Gorski RA, Gordon JH, Shryne JE, Southam AM. Evidence for a morphological sex difference within the medial preoptic area of the rat brain. Brain Res 1978; 148(2): 333-46.
[http://dx.doi.org/10.1016/0006-8993(78)90723-0] [PMID: 656937]

[138] Nishizuka M, Arai Y. Sexual dimorphism in synaptic organization in the amygdala and its dependence on neonatal hormone environment. Brain Res 1981; 212(1): 31-8.
[http://dx.doi.org/10.1016/0006-8993(81)90029-9] [PMID: 7225863]

[139] Breedlove SM, Arnold AP. Hormone accumulation in a sexually dimorphic motor nucleus of the rat spinal cord. Science 1980; 210(4469): 564-6.
[http://dx.doi.org/10.1126/science.7423210] [PMID: 7423210]

[140] Raskin K, de Gendt K, Duittoz A, *et al.* Conditional inactivation of androgen receptor gene in the nervous system: effects on male behavioral and neuroendocrine responses. J Neurosci 2009; 29(14): 4461-70.
[http://dx.doi.org/10.1523/JNEUROSCI.0296-09.2009] [PMID: 19357272]

[141] Juntti SA, Tollkuhn J, Wu MV, *et al.* The androgen receptor governs the execution, but not programming, of male sexual and territorial behaviors. Neuron 2010; 66(2): 260-72.
[http://dx.doi.org/10.1016/j.neuron.2010.03.024] [PMID: 20435002]

[142] Lange JD, Brown WA, Wincze JP, Zwick W. Serum testosterone concentration and penile tumescence changes in men. Horm Behav 1980; 14(3): 267-70.
[http://dx.doi.org/10.1016/0018-506X(80)90034-3] [PMID: 7429442]

[143] Obici S, Rossetti L. Minireview: nutrient sensing and the regulation of insulin action and energy balance. Endocrinology 2003; 144(12): 5172-8.
[http://dx.doi.org/10.1210/en.2003-0999] [PMID: 12970158]

[144] Schwartz MW, Porte D Jr. Diabetes, obesity, and the brain. Science 2005; 307(5708): 375-9.
[http://dx.doi.org/10.1126/science.1104344] [PMID: 15662002]

[145] Clegg DJ, Brown LM, Woods SC, Benoit SC. Gonadal hormones determine sensitivity to central leptin and insulin. Diabetes 2006; 55(4): 978-87.
[http://dx.doi.org/10.2337/diabetes.55.04.06.db05-1339] [PMID: 16567519]

[146] Yu IC, Lin HY, Liu NC, *et al.* Neuronal androgen receptor regulates insulin sensitivity *via* suppression of hypothalamic NF-κB-mediated PTP1B expression. Diabetes 2013; 62(2): 411-23.
[http://dx.doi.org/10.2337/db12-0135] [PMID: 23139353]

[147] Chang C, Yeh S, Lee SO, Chang T. Androgen receptor (AR) pathophysiological roles in androgen related diseases in skin, metabolism syndrome, bone/muscle and neuron/immune systems: lessons learned from mice lacking ar in specific cells. Nucl Recept Signal 2013; 11: e001.
[http://dx.doi.org/10.1621/nrs.11001] [PMID: 24653668]

[148] Kochakian CD. Mechanisms of androgen actions. Lab Invest 1959; 8(2): 538-56.
[PMID: 13642796]

[149] Muoio DM, Newgard CB. Obesity-related derangements in metabolic regulation. Annu Rev Biochem 2006; 75: 367-401.
[http://dx.doi.org/10.1146/annurev.biochem.75.103004.142512] [PMID: 16756496]

[150] Wu C, Okar DA, Kang J, Lange AJ. Reduction of hepatic glucose production as a therapeutic target in the treatment of diabetes. Curr Drug Targets Immune Endocr Metabol Disord 2005; 5(1): 51-9.
[http://dx.doi.org/10.2174/1568008053174769] [PMID: 15777204]

[151] Postic C, Dentin R, Girard J. Role of the liver in the control of carbohydrate and lipid homeostasis. Diabetes Metab 2004; 30(5): 398-408.
[http://dx.doi.org/10.1016/S1262-3636(07)70133-7] [PMID: 15671906]

[152] Samuel VT, Liu ZX, Qu X, *et al.* Mechanism of hepatic insulin resistance in non-alcoholic fatty liver disease. J Biol Chem 2004; 279(31): 32345-53.
[http://dx.doi.org/10.1074/jbc.M313478200] [PMID: 15166226]

[153] Lin HY, Yu IC, Wang RS, *et al.* Increased hepatic steatosis and insulin resistance in mice lacking hepatic androgen receptor. Hepatology 2008; 47(6): 1924-35.
[http://dx.doi.org/10.1002/hep.22252] [PMID: 18449947]

[154] Molinari PF, Rosenkrantz H. Erythropoietic activity and androgenic implications of 29 testosterone derivatives in orchiectomized rats. J Lab Clin Med 1971; 78(3): 399-410.
[PMID: 5092861]

[155] Bozzini CE, Alippi RM. The erythrogenic effect of steroids with predominant anabolic or androgenic activity in the polycythemic mouse. Horm Metab Res 1971; 3(1): 52-4.
[http://dx.doi.org/10.1055/s-0028-1095028] [PMID: 5128328]

[156] Vollmer EP, Gordon AS. Effect of sex and gonadotropic hormones upon the blood picture of the rat. Endocrinology 1941; 29: 828-37.
[http://dx.doi.org/10.1210/endo-29-5-828]

[157] Steinglass P, Gordon AS, Charipper HA. Effect of castration and sex hormones on blood of the rat. Proc Soc Exp Biol Med 1941; 48: 169-77.
[http://dx.doi.org/10.3181/00379727-48-13259]

[158] Crafts RC. Effects of hypophysectomy, castration, and testosterone propionate on hemopoiesis in the adult male rat. Endocrinology 1946; 39(6): 401-13.
[http://dx.doi.org/10.1210/endo-39-6-401] [PMID: 20282551]

[159] Hawkins WW, Speck E, Leonard VG. Variation of the hemoglobin level with age and sex. Blood 1954; 9(10): 999-1007.
[PMID: 13208753]

[160] Williamson CS. Influence of age and sex on hemoglobin: a spectrophotometric analysis of nine hundred and nineteen cases. Arch Intern Med (Chic) 1916; 18: 505-28.
[http://dx.doi.org/10.1001/archinte.1916.00080170078006]

[161] Vahlquist B. The cause of the sexual differences in erythrocyte hemoglobin and serum iron levels in human adults. Blood 1950; 5(9): 874-5.
[PMID: 15434015]

[162] McCullagh EP, Jones R. A note on the effect of certain androgens upon red blood cell count and upon glucose tolerance. Cleve Clin Q 1941; 8: 79-84.
[http://dx.doi.org/10.3949/ccjm.8.2.79]

[163] Kennedy BJ, Gilbertsen AS. Increased erythropoiesis induced by androgenic-hormone therapy. N Engl J Med 1957; 256(16): 719-26.
[http://dx.doi.org/10.1056/NEJM195704182561601] [PMID: 13451926]

[164] Nathan DG, Gardner FH. Effects of large doses of androgen on rodent erythropoiesis and body composition. Blood 1965; 26(4): 411-20.
[PMID: 5825006]

[165] Paulo LG, Fink GD, Roh BL, Fisher JW. Effects of several androgens and steroid metabolites on erythropoietin production in the isolated perfused dog kidney. Blood 1974; 43(1): 39-47.
[PMID: 4809096]

[166] Medlinsky JT, Napier CD, Gurney CW. The use of an antiandrogen to further investigate the erythropoietic effects of androgens. J Lab Clin Med 1969; 74(1): 85-92.
[PMID: 5789576]

[167] Moriyama Y, Fisher JW. Effects of testosterone and erythropoietin on erythroid colony formation in human bone marrow cultures. Blood 1975; 45(5): 665-70.
[PMID: 47251]

[168] Molinari PF, Rosenkrantz H. Erythropoietic activity and androgenic implications of 29 testosterone derivatives in orchiectomized rats. J Lab Clin Med 1971; 78(3): 399-410.
[PMID: 5092861]

[169] Valladares L, Minguell J. Characterization of a nuclear receptor for testosterone in rat bone marrow. Steroids 1975; 25(1): 13-21.
[http://dx.doi.org/10.1016/S0039-128X(75)80003-1] [PMID: 163042]

[170] Molinari PF, Esber HJ, Snyder LM. Effect of androgens on maturation and metabolism of erythroid tissue. Exp Hematol 1976; 4(5): 301-9.
[PMID: 976390]

[171] Larner J. Intermediary metabolism and its regulation. Englewood Cliffs, NJ: Prentice Hall Inc. 1971; p. 228.

[172] Molinari PF, Neri LL. Effect of a single oral dose of oxymetholone on the metabolism of human erythrocytes. Exp Hematol 1978; 6(8): 648-54.
[PMID: 710546]

[173] Necheles TF, Rai US. Studies on the control of hemoglobin synthesis: the *in vitro* stimulating effect of a 5-beta-H steroid metabolite on heme formation in human bone marrow cells. Blood 1969; 34(3): 380-4.
[PMID: 5804027]

[174] Irving RA, Mainwaring IP, Spooner PM. The regulation of haemoglobin synthesis in cultured chick blastoderms by steroids related to 5beta-androstane. Biochem J 1976; 154(1): 81-93.
[http://dx.doi.org/10.1042/bj1540081] [PMID: 1275915]

[175] Roubinian JR, Talal N, Greenspan JS, Goodman JR, Siiteri PK. Effect of castration and sex hormone treatment on survival, anti-nucleic acid antibodies, and glomerulonephritis in NZB/NZW F1 mice. J

Exp Med 1978; 147(6): 1568-83.
[http://dx.doi.org/10.1084/jem.147.6.1568] [PMID: 308087]

[176] Roubinian JR, Papoian R, Talal N. Androgenic hormones modulate autoantibody responses and improve survival in murine lupus. J Clin Invest 1977; 59(6): 1066-70.
[http://dx.doi.org/10.1172/JCI108729] [PMID: 864003]

[177] Ziehn MO, Avedisian AA, Dervin SM, Umeda EA, O'Dell TJ, Voskuhl RR. Therapeutic testosterone administration preserves excitatory synaptic transmission in the hippocampus during autoimmune demyelinating disease. J Neurosci 2012; 32(36): 12312-24.
[http://dx.doi.org/10.1523/JNEUROSCI.2796-12.2012] [PMID: 22956822]

[178] Pikwer M, Giwercman A, Bergström U, Nilsson JA, Jacobsson LT, Turesson C. Association between testosterone levels and risk of future rheumatoid arthritis in men: a population-based case-control study. Ann Rheum Dis 2014; 73(3): 573-9.
[http://dx.doi.org/10.1136/annrheumdis-2012-202781] [PMID: 23553100]

[179] Cook MB, Dawsey SM, Freedman ND, *et al.* Sex disparities in cancer incidence by period and age. Cancer Epidemiol Biomarkers Prev 2009; 18(4): 1174-82.
[http://dx.doi.org/10.1158/1055-9965.EPI-08-1118] [PMID: 19293308]

[180] Edgren G, Liang L, Adami HO, Chang ET. Enigmatic sex disparities in cancer incidence. Eur J Epidemiol 2012; 27(3): 187-96.
[http://dx.doi.org/10.1007/s10654-011-9647-5] [PMID: 22212865]

[181] Zhang LJ, Xiong Y, Nilubol N, *et al.* Testosterone regulates thyroid cancer progression by modifying tumor suppressor genes and tumor immunity. Carcinogenesis 2015; 36(4): 420-8.
[http://dx.doi.org/10.1093/carcin/bgv001] [PMID: 25576159]

[182] Amos-Landgraf JM, Heijmans J, Wielenga MC, *et al.* Sex disparity in colonic adenomagenesis involves promotion by male hormones, not protection by female hormones. Proc Natl Acad Sci USA 2014; 111(46): 16514-9.
[http://dx.doi.org/10.1073/pnas.1323064111] [PMID: 25368192]

[183] Byron JW. Effect of steroids on the cycling of haemopoietic stem cells. Nature 1970; 228(5277): 1204.
[http://dx.doi.org/10.1038/2281204a0] [PMID: 5487251]

[184] Udupa KB, Reissmann KR. Acceleration of granulopoietic recovery by androgenic steroids in mice made neutropenic by cytotoxic drugs. Cancer Res 1974; 34(10): 2517-20.
[PMID: 4411809]

[185] Horn Y, Price DC. The effect of androgens on erythroid and granulocytic marrow recovery in the irradiated rat. Acta Haematol 1972; 48(5): 300-6.
[http://dx.doi.org/10.1159/000208474] [PMID: 4629535]

[186] Sanchez-Medal L. The hemopoietic action of androstanes. Prog Hematol 1971; 7: 111-36.
[PMID: 4950580]

[187] Rawbone RG, Bagshawe KD. Anabolic steroids and bone marrow toxicity during therapy with methotrexate. Br J Cancer 1972; 26(5): 395-401.
[http://dx.doi.org/10.1038/bjc.1972.52] [PMID: 4343678]

[188] Ibáñez L, Jaramillo AM, Ferrer A, de Zegher F. High neutrophil count in girls and women with hyperinsulinaemic hyperandrogenism: normalization with metformin and flutamide overcomes the aggravation by oral contraception. Hum Reprod 2005; 20(9): 2457-62.
[http://dx.doi.org/10.1093/humrep/dei072] [PMID: 15905296]

[189] Chuang KH, Altuwaijri S, Li G, *et al.* Neutropenia with impaired host defense against microbial infection in mice lacking androgen receptor. J Exp Med 2009; 206(5): 1181-99.
[http://dx.doi.org/10.1084/jem.20082521] [PMID: 19414555]

[190] Trigunaite A, Khan A, Der E, Song A, Varikuti S, Jørgensen TN. Gr-1(high) CD11b+ cells suppress B cell differentiation and lupus-like disease in lupus-prone male mice. Arthritis Rheum 2013; 65(9):

2392-402.
[http://dx.doi.org/10.1002/art.38048] [PMID: 23754362]

[191] Arango Duque G, Descoteaux A. Macrophage cytokines: involvement in immunity and infectious diseases. Front Immunol 2014; 5: 491.
[http://dx.doi.org/10.3389/fimmu.2014.00491] [PMID: 25339958]

[192] Sieweke MH, Allen JE. Beyond stem cells: self-renewal of differentiated macrophages. Science 2013; 342(6161): 1242974.
[http://dx.doi.org/10.1126/science.1242974] [PMID: 24264994]

[193] Takahashi K. Development and differentiation of macrophages and related cells: historical review and current concept. J Clin Exp Hematop 2001; 41: 1-33.
[http://dx.doi.org/10.3960/jslrt.41.1]

[194] Lai JJ, Lai KP, Chuang KH, *et al.* Monocyte/macrophage androgen receptor suppresses cutaneous wound healing in mice by enhancing local TNF-alpha expression. J Clin Invest 2009; 119(12): 3739-51.
[http://dx.doi.org/10.1172/JCI39335] [PMID: 19907077]

[195] Viselli SM, Reese KR, Fan J, Kovacs WJ, Olsen NJ. Androgens alter B cell development in normal male mice. Cell Immunol 1997; 182(2): 99-104.
[http://dx.doi.org/10.1006/cimm.1997.1227] [PMID: 9514700]

[196] Smithson G, Beamer WG, Shultz KL, Christianson SW, Shultz LD, Kincade PW. Increased B lymphopoiesis in genetically sex steroid-deficient hypogonadal (hpg) mice. J Exp Med 1994; 180(2): 717-20.
[http://dx.doi.org/10.1084/jem.180.2.717] [PMID: 8046347]

[197] Ellis TM, Moser MT, Le PT, Flanigan RC, Kwon ED. Alterations in peripheral B cells and B cell progenitors following androgen ablation in mice. Int Immunol 2001; 13(4): 553-8.
[http://dx.doi.org/10.1093/intimm/13.4.553] [PMID: 11282994]

[198] Altuwaijri S, Chuang KH, Lai KP, *et al.* Susceptibility to autoimmunity and B cell resistance to apoptosis in mice lacking androgen receptor in B cells. Mol Endocrinol 2009; 23(4): 444-53.
[http://dx.doi.org/10.1210/me.2008-0106] [PMID: 19164450]

[199] Montes CL, Maletto BA, Acosta Rodriguez EV, Gruppi A, Pistoresi-Palencia MC. B cells from aged mice exhibit reduced apoptosis upon B-cell antigen receptor stimulation and differential ability to up-regulate survival signals. Clin Exp Immunol 2006; 143(1): 30-40.
[http://dx.doi.org/10.1111/j.1365-2249.2005.02969.x] [PMID: 16367931]

[200] Olsen NJ, Gu X, Kovacs WJ. Bone marrow stromal cells mediate androgenic suppression of B lymphocyte development. J Clin Invest 2001; 108(11): 1697-704.
[http://dx.doi.org/10.1172/JCI200113183] [PMID: 11733565]

[201] Tang J, Nuccie BL, Ritterman I, Liesveld JL, Abboud CN, Ryan DH. TGF-beta down-regulates stromal IL-7 secretion and inhibits proliferation of human B cell precursors. J Immunol 1997; 159(1): 117-25.
[PMID: 9200446]

[202] Kumar N, Shan LX, Hardy MP, Bardin CW, Sundaram K. Mechanism of androgen-induced thymolysis in rats. Endocrinology 1995; 136(11): 4887-93.
[http://dx.doi.org/10.1210/endo.136.11.7588221] [PMID: 7588221]

[203] Sutherland JS, Goldberg GL, Hammett MV, *et al.* Activation of thymic regeneration in mice and humans following androgen blockade. J Immunol 2005; 175(4): 2741-53.
[http://dx.doi.org/10.4049/jimmunol.175.4.2741] [PMID: 16081852]

[204] Olsen NJ, Olson G, Viselli SM, Gu X, Kovacs WJ. Androgen receptors in thymic epithelium modulate thymus size and thymocyte development. Endocrinology 2001; 142(3): 1278-83.
[http://dx.doi.org/10.1210/endo.142.3.8032] [PMID: 11181545]

[205] Heng TS, Goldberg GL, Gray DH, Sutherland JS, Chidgey AP, Boyd RL. Effects of castration on thymocyte development in two different models of thymic involution. J Immunol 2005; 175(5): 2982-93.
[http://dx.doi.org/10.4049/jimmunol.175.5.2982] [PMID: 16116185]

[206] Olsen NJ, Kovacs WJ. Evidence that androgens modulate human thymic T cell output. J Investig Med 2011; 59(1): 32-5.
[http://dx.doi.org/10.2310/JIM.0b013e318200dc98] [PMID: 21218609]

[207] LeMaoult J, Messaoudi I, Manavalan JS, *et al*. Age-related dysregulation in CD8 T cell homeostasis: kinetics of a diversity loss. J Immunol 2000; 165(5): 2367-73.
[http://dx.doi.org/10.4049/jimmunol.165.5.2367] [PMID: 10946259]

[208] Haynes L, Maue AC. Effects of aging on T cell function. Curr Opin Immunol 2009; 21(4): 414-7.
[http://dx.doi.org/10.1016/j.coi.2009.05.009] [PMID: 19500967]

[209] Trigunaite A, Dimo J, Jørgensen TN. Suppressive effects of androgens on the immune system. Cell Immunol 2015; 294(2): 87-94.
[http://dx.doi.org/10.1016/j.cellimm.2015.02.004] [PMID: 25708485]

[210] Basaria S. Androgen abuse in athletes: detection and consequences. J Clin Endocrinol Metab 2010; 95(4): 1533-43.
[http://dx.doi.org/10.1210/jc.2009-1579] [PMID: 20139230]

[211] Ikeda Y, Aihara K, Sato T, *et al*. Androgen receptor gene knockout male mice exhibit impaired cardiac growth and exacerbation of angiotensin II-induced cardiac fibrosis. J Biol Chem 2005; 280(33): 29661-6.
[http://dx.doi.org/10.1074/jbc.M411694200] [PMID: 15961403]

[212] Nettleship JE, Jones TH, Channer KS, Jones RD. Physiological testosterone replacement therapy attenuates fatty streak formation and improves high-density lipoprotein cholesterol in the Tfm mouse: an effect that is independent of the classic androgen receptor. Circulation 2007; 116(21): 2427-34.
[http://dx.doi.org/10.1161/CIRCULATIONAHA.107.708768] [PMID: 17984376]

[213] Ikeda Y, Aihara K, Yoshida S, *et al*. Androgen-androgen receptor system protects against angiotensin II-induced vascular remodeling. Endocrinology 2009; 150(6): 2857-64.
[http://dx.doi.org/10.1210/en.2008-1254] [PMID: 19196803]

[214] Huang CK, Lee SO, Chang C. The male hormone signals in cardiovascular diseases J Hypertension 2012; 30 (Suppl 1) e311 Abstract #1069

[215] Vanderschueren D, Vandenput L, Boonen S, Lindberg MK, Bouillon R, Ohlsson C. Androgens and bone. Endocr Rev 2004; 25(3): 389-425.
[http://dx.doi.org/10.1210/er.2003-0003] [PMID: 15180950]

[216] Basaria S, Dobs AS. Hypogonadism and androgen replacement therapy in elderly men. Am J Med 2001; 110(7): 563-72.
[http://dx.doi.org/10.1016/S0002-9343(01)00663-5] [PMID: 11343670]

[217] Wiren KM. Androgens and bone growth: it's location, location, location. Curr Opin Pharmacol 2005; 5(6): 626-32.
[http://dx.doi.org/10.1016/j.coph.2005.06.003] [PMID: 16185926]

[218] MacLean HE, Moore AJ, Sastra SA, *et al*. DNA-binding-dependent androgen receptor signaling contributes to gender differences and has physiological actions in males and females. J Endocrinol 2010; 206(1): 93-103.
[http://dx.doi.org/10.1677/JOE-10-0026] [PMID: 20395380]

[219] Kawano H, Sato T, Yamada T, *et al*. Suppressive function of androgen receptor in bone resorption. Proc Natl Acad Sci USA 2003; 100(16): 9416-21.
[http://dx.doi.org/10.1073/pnas.1533500100] [PMID: 12872002]

[220] Yeh S, Tsai MY, Xu Q, *et al.* Generation and characterization of androgen receptor knockout (ARKO) mice: an *in vivo* model for the study of androgen functions in selective tissues. Proc Natl Acad Sci USA 2002; 99(21): 13498-503.
[http://dx.doi.org/10.1073/pnas.212474399] [PMID: 12370412]

[221] Tsai MY, Shyr CR, Kang HY, *et al.* The reduced trabecular bone mass of adult ARKO male mice results from the decreased osteogenic differentiation of bone marrow stroma cells. Biochem Biophys Res Commun 2011; 411(3): 477-82.
[http://dx.doi.org/10.1016/j.bbrc.2011.06.113] [PMID: 21723262]

[222] Frenkel B, Hong A, Baniwal SK, *et al.* Regulation of adult bone turnover by sex steroids. J Cell Physiol 2010; 224(2): 305-10.
[http://dx.doi.org/10.1002/jcp.22159] [PMID: 20432458]

[223] Colvard DS, Eriksen EF, Keeting PE, *et al.* Identification of androgen receptors in normal human osteoblast-like cells. Proc Natl Acad Sci USA 1989; 86(3): 854-7.
[http://dx.doi.org/10.1073/pnas.86.3.854] [PMID: 2915981]

[224] Kasperk CH, Wergedal JE, Farley JR, Linkhart TA, Turner RT, Baylink DJ. Androgens directly stimulate proliferation of bone cells *in vitro.* Endocrinology 1989; 124(3): 1576-8.
[http://dx.doi.org/10.1210/endo-124-3-1576] [PMID: 2521824]

[225] Wiren KM, Zhang XW, Toombs AR, *et al.* Targeted overexpression of androgen receptor in osteoblasts: unexpected complex bone phenotype in growing animals. Endocrinology 2004; 145(7): 3507-22.
[http://dx.doi.org/10.1210/en.2003-1016] [PMID: 15131013]

[226] Wiren KM, Semirale AA, Zhang XW, *et al.* Targeting of androgen receptor in bone reveals a lack of androgen anabolic action and inhibition of osteogenesis: a model for compartment-specific androgen action in the skeleton. Bone 2008; 43(3): 440-51.
[http://dx.doi.org/10.1016/j.bone.2008.04.026] [PMID: 18595795]

[227] Zhang M, Xuan S, Bouxsein ML, *et al.* Osteoblast-specific knockout of the insulin-like growth factor (IGF) receptor gene reveals an essential role of IGF signaling in bone matrix mineralization. J Biol Chem 2002; 277(46): 44005-12.
[http://dx.doi.org/10.1074/jbc.M208265200] [PMID: 12215457]

[228] Notini AJ, McManus JF, Moore A, *et al.* Osteoblast deletion of exon 3 of the androgen receptor gene results in trabecular bone loss in adult male mice. J Bone Miner Res 2007; 22(3): 347-56.
[http://dx.doi.org/10.1359/jbmr.061117] [PMID: 17147488]

[229] Chiang C, Chiu M, Moore AJ, *et al.* Mineralization and bone resorption are regulated by the androgen receptor in male mice. J Bone Miner Res 2009; 24(4): 621-31.
[http://dx.doi.org/10.1359/jbmr.081217] [PMID: 19049333]

[230] Sinnesael M, Claessens F, Laurent M, *et al.* Androgen receptor (AR) in osteocytes is important for the maintenance of male skeletal integrity: evidence from targeted AR disruption in mouse osteocytes. J Bone Miner Res 2012; 27(12): 2535-43.
[http://dx.doi.org/10.1002/jbmr.1713] [PMID: 22836391]

[231] Celotti F, Negri Cesi P. Anabolic steroids: a review of their effects on the muscles, of their possible mechanisms of action and of their use in athletics. J Steroid Biochem Mol Biol 1992; 43(5): 469-77.
[http://dx.doi.org/10.1016/0960-0760(92)90085-W] [PMID: 1390296]

[232] Chen Y, Zajac JD, MacLean HE. Androgen regulation of satellite cell function. J Endocrinol 2005; 186(1): 21-31.
[http://dx.doi.org/10.1677/joe.1.05976] [PMID: 16002532]

[233] Bhasin S, Calof OM, Storer TW, *et al.* Drug insight: Testosterone and selective androgen receptor modulators as anabolic therapies for chronic illness and aging. Nat Clin Pract Endocrinol Metab 2006; 2(3): 146-59.

[http://dx.doi.org/10.1038/ncpendmet0120] [PMID: 16932274]

[234] Herbst KL, Bhasin S. Testosterone action on skeletal muscle. Curr Opin Clin Nutr Metab Care 2004; 7(3): 271-7.
[http://dx.doi.org/10.1097/00075197-200405000-00006] [PMID: 15075918]

[235] Sinha-Hikim I, Artaza J, Woodhouse L, *et al.* Testosterone-induced increase in muscle size in healthy young men is associated with muscle fiber hypertrophy. Am J Physiol Endocrinol Metab 2002; 283(1): E154-64.
[http://dx.doi.org/10.1152/ajpendo.00502.2001] [PMID: 12067856]

[236] Sinha-Hikim I, Roth SM, Lee MI, Bhasin S. Testosterone-induced muscle hypertrophy is associated with an increase in satellite cell number in healthy, young men. Am J Physiol Endocrinol Metab 2003; 285(1): E197-205.
[http://dx.doi.org/10.1152/ajpendo.00370.2002] [PMID: 12670837]

[237] Sinha-Hikim I, Cornford M, Gaytan H, Lee ML, Bhasin S. Effects of testosterone supplementation on skeletal muscle fiber hypertrophy and satellite cells in community-dwelling older men. J Clin Endocrinol Metab 2006; 91(8): 3024-33.
[http://dx.doi.org/10.1210/jc.2006-0357] [PMID: 16705073]

[238] Ting HJ, Chang C. Actin associated proteins function as androgen receptor coregulators: an implication of androgen receptor's roles in skeletal muscle. J Steroid Biochem Mol Biol 2008; 111(3-5): 157-63.
[http://dx.doi.org/10.1016/j.jsbmb.2008.06.001] [PMID: 18590822]

[239] Altuwaijri S, Lee DK, Chuang KH, *et al.* Androgen receptor regulates expression of skeletal muscle-specific proteins and muscle cell types. Endocrine 2004; 25(1): 27-32.
[http://dx.doi.org/10.1385/ENDO:25:1:27] [PMID: 15545703]

[240] Brown TJ, Khan T, Jones KJ. Androgen induced acceleration of functional recovery after rat sciatic nerve injury. Restor Neurol Neurosci 1999; 15(4): 289-95.
[PMID: 12671219]

[241] Boissonneault G. Evidence of apoptosis in the castration-induced atrophy of the rat levator ani muscle. Endocr Res 2001; 27(3): 317-28.
[http://dx.doi.org/10.1081/ERC-100106009] [PMID: 11678579]

[242] Callewaert F, Venken K, Ophoff J, *et al.* Differential regulation of bone and body composition in male mice with combined inactivation of androgen and estrogen receptor-alpha. FASEB J 2009; 23(1): 232-40.
[http://dx.doi.org/10.1096/fj.08-113456] [PMID: 18809737]

[243] Ophoff J, Van Proeyen K, Callewaert F, *et al.* Androgen signaling in myocytes contributes to the maintenance of muscle mass and fiber type regulation but not to muscle strength or fatigue. Endocrinology 2009; 150(8): 3558-66.
[http://dx.doi.org/10.1210/en.2008-1509] [PMID: 19264874]

[244] Srinivas-Shankar U, Roberts SA, Connolly MJ, *et al.* Effects of testosterone on muscle strength, physical function, body composition, and quality of life in intermediate-frail and frail elderly men: a randomized, double-blind, placebo-controlled study. J Clin Endocrinol Metab 2010; 95(2): 639-50.
[http://dx.doi.org/10.1210/jc.2009-1251] [PMID: 20061435]

[245] Mauras N, Hayes V, Welch S, *et al.* Testosterone deficiency in young men: marked alterations in whole body protein kinetics, strength, and adiposity. J Clin Endocrinol Metab 1998; 83(6): 1886-92.
[PMID: 9626114]

[246] Basaria S, Lieb J II, Tang AM, *et al.* Long-term effects of androgen deprivation therapy in prostate cancer patients. Clin Endocrinol (Oxf) 2002; 56(6): 779-86.
[http://dx.doi.org/10.1046/j.1365-2265.2002.01551.x] [PMID: 12072048]

[247] Lexell J. Ageing and human muscle: observations from Sweden. Can J Appl Physiol 1993; 18(1): 2-

18.
[http://dx.doi.org/10.1139/h93-002] [PMID: 8471991]

[248] Baumgartner RN, Waters DL, Gallagher D, Morley JE, Garry PJ. Predictors of skeletal muscle mass in elderly men and women. Mech Ageing Dev 1999; 107(2): 123-36.
[http://dx.doi.org/10.1016/S0047-6374(98)00130-4] [PMID: 10220041]

[249] Roubenoff R, Hughes VA. Sarcopenia: current concepts. J Gerontol A Biol Sci Med Sci 2000; 55(12): M716-24.
[http://dx.doi.org/10.1093/gerona/55.12.M716] [PMID: 11129393]

[250] Dirks A, Leeuwenburgh C. Apoptosis in skeletal muscle with aging. Am J Physiol Regul Integr Comp Physiol 2002; 282(2): R519-27.
[http://dx.doi.org/10.1152/ajpregu.00458.2001] [PMID: 11792662]

[251] Dupont-Versteegden EE. Apoptosis in muscle atrophy: relevance to sarcopenia. Exp Gerontol 2005; 40(6): 473-81.
[http://dx.doi.org/10.1016/j.exger.2005.04.003] [PMID: 15935591]

[252] Pronsato L, Ronda AC, Milanesi LM, Vasconsuelo AA, Boland RL. Protective role of 17β-estradiol and testosterone in apoptosis of skeletal muscle. Actual Osteol 2010; 6: 66-80.

[253] Schaible TF, Malhotra A, Ciambrone G, Scheuer J. The effects of gonadectomy on left ventricular function and cardiac contractile proteins in male and female rats. Circ Res 1984; 54(1): 38-49.
[http://dx.doi.org/10.1161/01.RES.54.1.38] [PMID: 6229365]

[254] Gilliver SC, Wu F, Ashcroft GS. Regulatory roles of androgens in cutaneous wound healing. Thromb Haemost 2003; 90(6): 978-85.
[http://dx.doi.org/10.1160/TH03-05-0302] [PMID: 14652627]

[255] Markova MS, Zeskand J, McEntee B, Rothstein J, Jimenez SA, Siracusa LD. A role for the androgen receptor in collagen content of the skin. J Invest Dermatol 2004; 123(6): 1052-6.
[http://dx.doi.org/10.1111/j.0022-202X.2004.23494.x] [PMID: 15610513]

[256] Hamilton JB. Male hormone stimulation is prerequisite and an incitant in common baldness. Am J Anat 1942; 71: 451-80.
[http://dx.doi.org/10.1002/aja.1000710306]

[257] Rittmaster RS. Hirsutism. Lancet 1997; 349(9046): 191-5.
[http://dx.doi.org/10.1016/S0140-6736(96)07252-2] [PMID: 9111556]

[258] Imperato-McGinley J, Gautier T, Cai LQ, Yee B, Epstein J, Pochi P. The androgen control of sebum production. Studies of subjects with dihydrotestosterone deficiency and complete androgen insensitivity. J Clin Endocrinol Metab 1993; 76(2): 524-8.
[PMID: 8381804]

[259] Inui S, Fukuzato Y, Nakajima T, Yoshikawa K, Itami S. Androgen-inducible TGF-beta1 from balding dermal papilla cells inhibits epithelial cell growth: a clue to understand paradoxical effects of androgen on human hair growth. FASEB J 2002; 16(14): 1967-9.
[http://dx.doi.org/10.1096/fj.02-0043fje] [PMID: 12397096]

[260] Davis BK. Quantitative morphological studies upon the influence of the endocrine system on the growth of hair by white mice. Acta Endocrinol (Copenh) 1963; 44 (Suppl. 85): L85-, 1-102.
[PMID: 14068350]

[261] Ebling FJ, Hale PA. The control of the mammalian moult. Mem Soc Endocrinol 1970; 18: 215-37.

[262] Naito A, Sato T, Matsumoto T, *et al.* Dihydrotestosterone inhibits murine hair growth *via* the androgen receptor. Br J Dermatol 2008; 159(2): 300-5.
[http://dx.doi.org/10.1111/j.1365-2133.2008.08671.x] [PMID: 18547308]

[263] Ashcroft GS, Mills SJ. Androgen receptor-mediated inhibition of cutaneous wound healing. J Clin Invest 2002; 110(5): 615-24.

[http://dx.doi.org/10.1172/JCI0215704] [PMID: 12208862]

[264] Gilliver SC, Ruckshanthi JP, Hardman MJ, Nakayama T, Ashcroft GS. Sex dimorphism in wound healing: the roles of sex steroids and macrophage migration inhibitory factor. Endocrinology 2008; 149(11): 5747-57.
[http://dx.doi.org/10.1210/en.2008-0355] [PMID: 18653719]

[265] Lai JJ, Lai KP, Chuang KH, *et al.* Monocyte/macrophage androgen receptor suppresses cutaneous wound healing in mice by enhancing local TNF-alpha expression. J Clin Invest 2009; 119(12): 3739-51.
[http://dx.doi.org/10.1172/JCI39335] [PMID: 19907077]

[266] Secchi J, Lecaque D, Bouton MM. Morphometric evaluation of local anti-androgenic activity on the rat sebaceous gland. Br J Dermatol 1984; 111 (Suppl. 27): 178-9.
[http://dx.doi.org/10.1111/j.1365-2133.1984.tb15601.x] [PMID: 6234921]

[267] Evans DJ, Hoffmann RG, Kalkhoff RK, Kissebah AH. Relationship of androgenic activity to body fat topography, fat cell morphology, and metabolic aberrations in premenopausal women. J Clin Endocrinol Metab 1983; 57(2): 304-10.
[http://dx.doi.org/10.1210/jcem-57-2-304] [PMID: 6345569]

[268] Lin HY, Xu Q, Yeh S, Wang RS, Sparks JD, Chang C. Insulin and leptin resistance with hyperleptinemia in mice lacking androgen receptor. Diabetes 2005; 54(6): 1717-25.
[http://dx.doi.org/10.2337/diabetes.54.6.1717] [PMID: 15919793]

[269] Dati E, Baroncelli GI, Mora S, *et al.* Body composition and metabolic profile in women with complete androgen insensitivity syndrome. Sex Dev 2009; 3(4): 188-93.
[http://dx.doi.org/10.1159/000228719] [PMID: 19752598]

[270] Yu IC, Lin HY, Liu NC, *et al.* Hyperleptinemia without obesity in male mice lacking androgen receptor in adipose tissue. Endocrinology 2008; 149(5): 2361-8.
[http://dx.doi.org/10.1210/en.2007-0516] [PMID: 18276764]

[271] McInnes KJ, Smith LB, Hunger NI, Saunders PT, Andrew R, Walker BR. Deletion of the androgen receptor in adipose tissue in male mice elevates retinol binding protein 4 and reveals independent effects on visceral fat mass and on glucose homeostasis. Diabetes 2012; 61(5): 1072-81.
[http://dx.doi.org/10.2337/db11-1136] [PMID: 22415878]

CHAPTER 2

Subcellular Localization Of Estrogen Receptors

Lorena M. Milanesi[*]

Instituto de Ciencias Biológicas y Biomédicas de Sur (INBIOSUR)-UNS-CONICET, Bahía Blanca, Argentina

Abstract: It is well established that estrogens elicit a variety of rapid effects in many tissues in addition to their actions on gene expression in the cell nucleus after Estrogen Receptors (ERs) dimerization and DNA-binding through their EREs. In this chapter we describe tissue distribution, subcellular localization, targets and roles of classical and non-classical ERs.

Keywords: Estrogen Receptor, ER α and β, Golgi, Mitochondria, Subcellular Localization.

INTRODUCTION

Estrogen Receptors

Estrogen receptor exists as two isoforms: ER α and β (Fig. **1**). Both are products of distinct genes on different chromosomes. ERα is located at chromosomal locus 6q25.1 [1], whereas ERβ is found in chromosome 14, position 14q22–24 [2]. Several splice variants for both ER have been described, but not all transcripts are expressed as functional proteins and have biological functions or it remains unclear [3].

In view of the multiple ER mRNA splice variants that exist and the great variety of functional proteins for some of these ERs, this could contribute to another level of complexity for the estrogen signaling/functions.

Tissue Distribution of ERs

The two ERs isoforms are widely distributed throughout the body [4, 5]. ERα is predominantly expressed in the uterus, kidney, heart and liver. ERβ is expressed not only in the ovary but also in prostate, lung, bladder, gastrointestinal tract, and hematopoietic and central nervous systems. Also ERs are coexpressed in several

[*] **Corresponding author Lorena M. Milanesi:** Instituto de Ciencias Biológicas y Biomédicas de Sur (INBIOSUR)-UNS-CONICET; Tel: +54 291 4595101x4337; E-mail: milanesi@criba.edu.ar

Andrea Vasconsuelo (Ed.)

tissues including the mammary gland, adrenal, thyroid, bone, epididymis, and certain regions of the brain. ER subtypes expressed in the same tissue can also be expressed in the same cell type and are targets for heterodimerization [3].

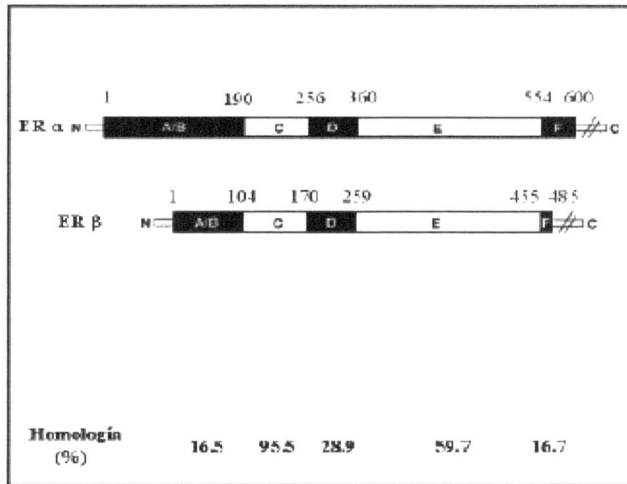

Fig. (1). ER α and β homology, structure and functions. **A/B domain:**with the lowest homology degree between ER α and β contains the activation function domain 1, AF-1, a transactivation domain, ligand independent. **C domain:**is the most conserved region between the two receptors (97%) and contains de DNA binding domain that consists of two zinc fingers motifs and various subdomains like a P-box (DNA binding specificity) and a D-box (dimerization interface). **D domain**: a low conserved domain with a hinge region and a heat shock protein (HSP) binding site. **E/F domain:**highly conserved between the two receptors. It is a multifunctional region with ligand binding domain, agonist and antagonist binding domains, dimerization domain, HSP and cofactors binding domains, nuclear signal localization and an activation function domain, ligand dependent (AF-2). Extracted from *Current Opinion in Pharmacology.*

Non Classical ERs and Non-Classical Localization

The classical model of 17-estradiol (E2) action has been traditionally described to be mediated by the estrogen receptor (ER) localized exclusively in the nucleus. However, there is increasing functional evidence for extra nuclear localization of ER.

Estrogen receptors exist in discrete cellular locations, including the nucleus and extra-nuclear compartments [6]. Both ER isoforms and their splice variants were detected in the plasma membrane and organelles such as mitochondria and the endoplasmic reticulum. This is known as non-classical ERs, but their role for some of them still not well established.

Several data favor the idea that some of the membrane-localized receptors are the same proteins as the nuclear ER receptors, transported to the plasma membrane by still unknown mechanisms. This concept is supported by immunocytochemistry studies [7 - 9], loss of immunodetection of membrane ERs using antisense oligonucleotides directed ERα [10] and codetection of membrane and nuclear ERs after nuclear ER cDNA expression in ER null cells [11]. However, investigations did not detect endogenous membrane and nuclear ERs in cells from DERKO mouse (ERα and ERβ deleted) by Western blot analysis [12].

On the other hand, there is evidence of the presence of estrogen binding proteins, possibly ion channels or GPCR like proteins in membrane subcellular compartments, not structurally but functionally related to ERs [13].

Some motifs of the classical ERs are critical for their membrane localization, dimerization and function [12, 14, 15]. Ser 522 of ERα has been shown to be necessary for the efficient translocation of ERα at the plasma membrane [16].

Cystine 447 has been shown to be crucial for steroid-independent palmitoylation of the receptor [17, 18]. This amino acid residue localized in the E domain is necessary for ERα to physically associate with the caveolin-1 protein and localize at the membrane. Also, palmitoylation has been proposed as a possible mechanism for membrane ER localization in endothelial cells [19].

Membrane ERs possibly exist as a pool attached to the inner face of the plasma membrane or associated with membrane proteins such as caveolin-1 [12], in conjunction with MNAR (modulator of nongenomic action of estrogen receptor) [20], or Shc and IGF-I [21].

It has been demonstrated that ERs can physically interact with G protein subunits, directly or indirectly (by binding to membrane proteins like caveolin-1) [11, 12, 22, 24] and it has been confirmed the ability of ERs to activate G protein-related signaling at the plasma membrane [11, 22 - 24].

In C2C12 skeletal muscle cell line, immunological and molecular data support mitochondrial-microsomal (Golgi) localization of ER α and β in the C2C12 (Figs. **2** and **3**) [25, 26]. Binding characteristics in whole cells in culture were established and immunocytological studies revealed that estrogen receptors mainly localize at the mitochondrial and perinuclear level.

These results were also confirmed using fluorescent 17β−estradiol-BSA conjugates. The immunoreactivity did not translocate into the nucleus by E2 treatment. Western and Ligand blot approaches corroborated the non-classical localization. Expression and subcellular distribution of ER α proteins were confirmed in C2C12 cells transfected with ER α siRNA and by RT-PCR employing specific primers (Figs. **4** and **5**).

TE111.5D11

AER 314

AER 308

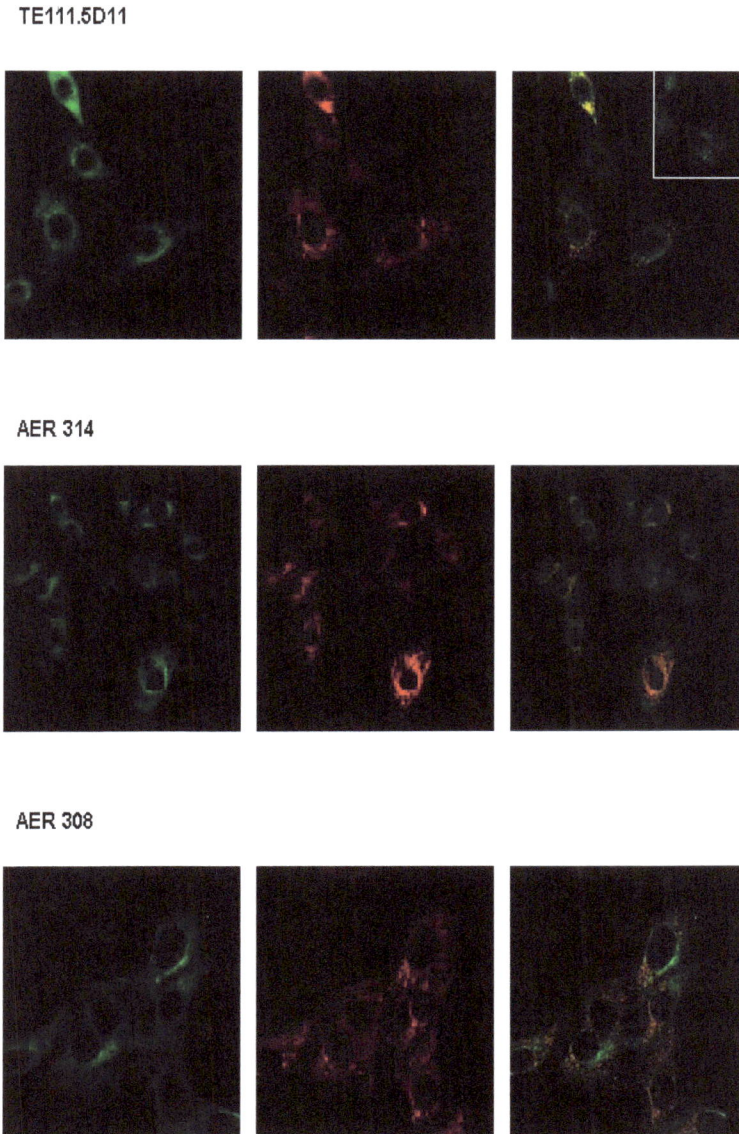

Fig. (2). Immunofluorescence confocal microscopy of C2C12 muscle cells after staining for ER α antigen. A. Whole cells. C2C12 cells were labeled using TE111.5D11, AER 314 and AER 308 antibodies. ER α specific green fluorescence was bright in mitochondria and perinuclear compartment of the cells (left). Mitochondria localization in C2C12 cells with the marker MitoTracker rendered a red signal (center) and the merged image of ER α immunostaining and the mitochondrial marker gave an orange/yellow signal (right). The merged image at the corner show the presence of punctuates nuclear fluorescence in some cells when the TE111.5D11 antibody was employed. From: Milanesi *et al.*, 2008.

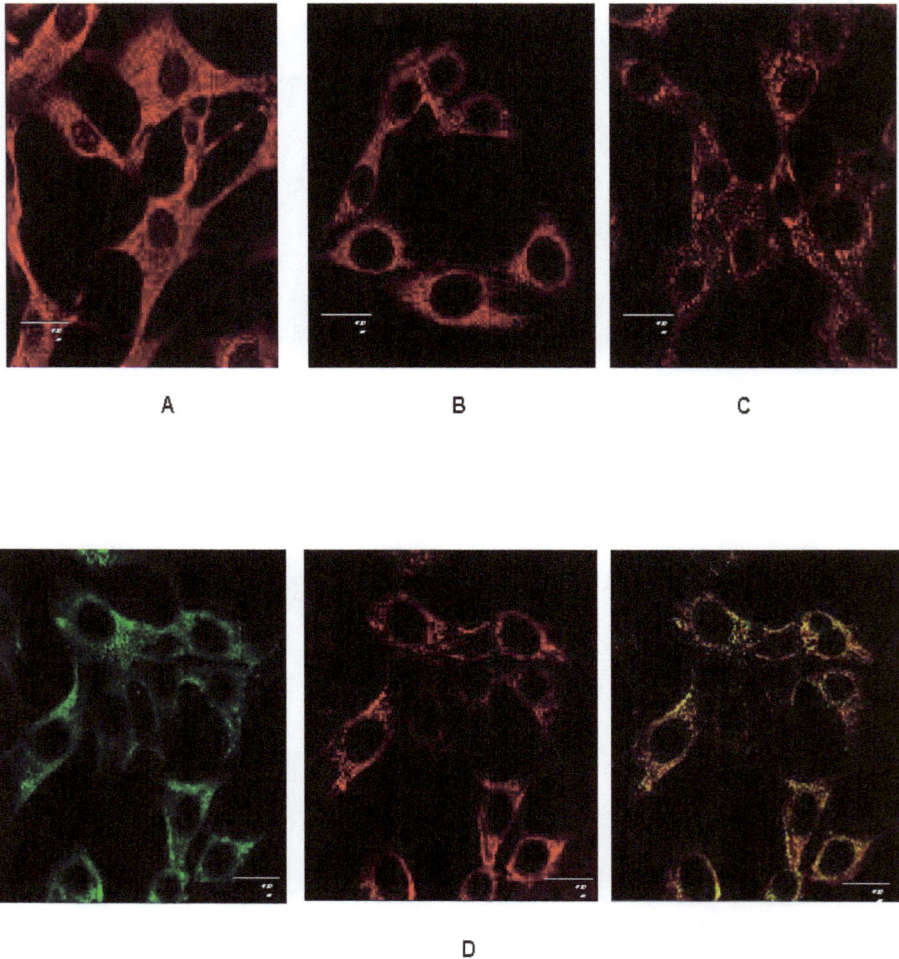

Fig. (3). Immunofluorescence confocal microscopy s of C2C12 muscle cells after staining for ER β antigen. C2C12 cells were labeled as described in Experimental, using polyclonal antibodies against ER β, epitope specific **(A)**, Y19 **(B)** L20 **(C)** and H-150 **(D)**. ER β specific green fluorescence was bright in mitochondrial cell compartments. Staining of mitochondria with the specific marker MitoTracker Red rendered a red signal (D, center) and yellow for the merged image of ER β and mitochondria (D, right). From: Milanesi *et al.*, 2009.

The non-classical distribution of native pools of ERs in skeletal muscle cells suggests an alternative mode of ER localization/function.

These interactions of non-nuclear ERs with cytosolic or membrane proteins can regulate gene transcription, independently of nuclear ERs. In addition, the membrane ER mediates estrogen-induced posttranslational modifications of several existing proteins [27].

Control

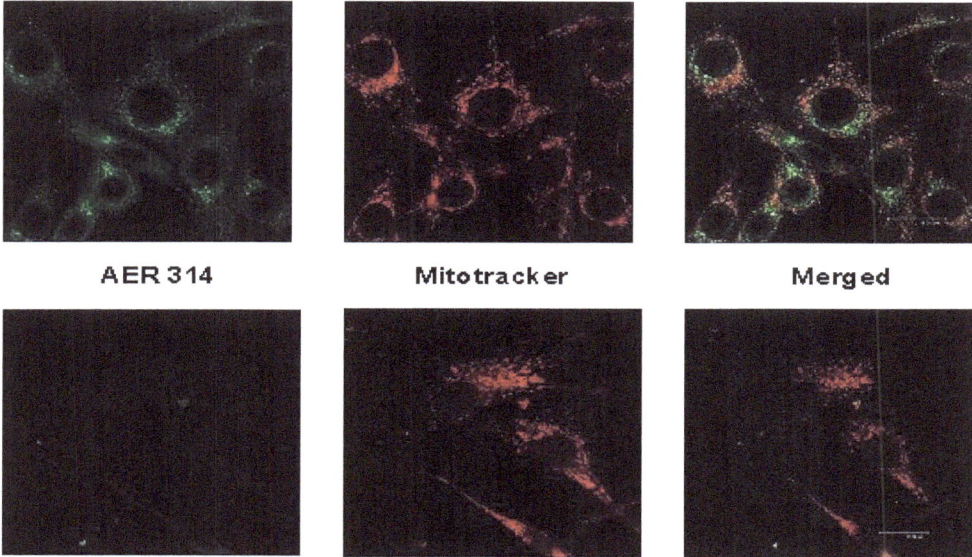

Fig. (4). Effects of transfection of ER α siRNA on ER α immunoreactivity. C2C12 cells were transiently transfected with siRNA probes followed by fluorescence microscopy. Original magnification= 400 X. B. Cells were transiently transfected with 20 pmol of ER α siRNA, incubated 24 h and then fixed and labeled for immunocytochemistry. Upper panel: Control (non-transfected cells). Left: ER α specific green fluorescence in nucleus, perinuclear region and cytosol employing AER 314 as primary antibody. From: Milanesi *et al.*, 2008.

Thus, estradiol can act through classical ERs or other receptors like splice variants or estrogen binding proteins localized in or near the cell membrane, the interaction can signal to the nucleus through multiple kinases altering the transcriptional activity of nuclear ERs and/or other transcription factors. The ability of these proteins to signal through multiple kinases to the nucleus, impacts several aspects of cellular function.

CONCLUDING REMARKS

The presence not only of the classical nuclear ER but also of membrane and mitochondrial as well as other proteins able to bind estradiol and to trigger signaling responses, has changed our thinking about hormone mechanism and their action. The existence of interaction between ERs and other receptor or non-receptor proteins in different subcellular compartments demonstrated the fine

Control

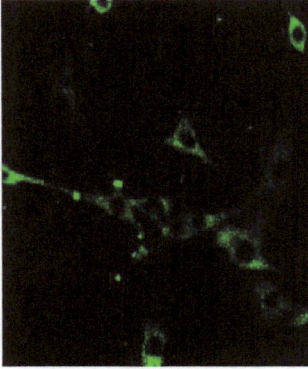 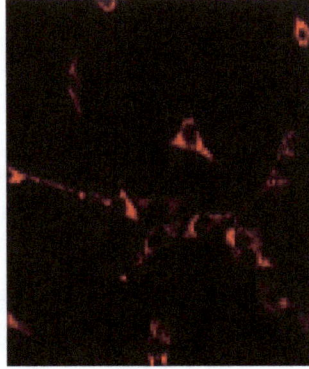 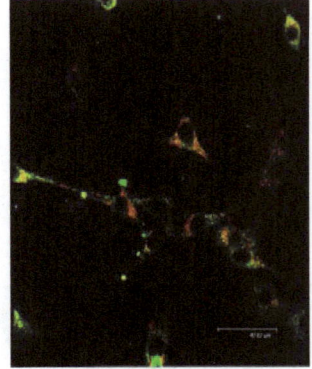

anti-ER beta: H-150   **Mitotracker**   **Merged**

 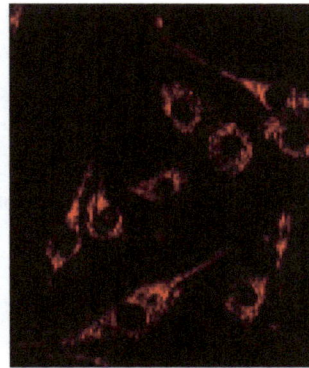 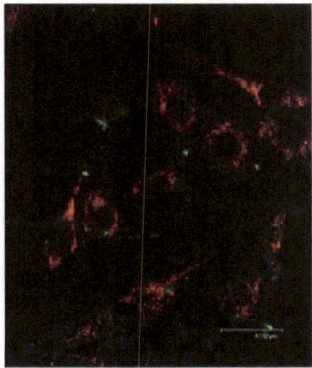

ERβ ShortCut siRNA

Fig. (5). Effects of transfection of ER β siRNA on ER β immunoreactivity. C2C12 cells were transiently transfected with siRNA probes followed by fluorescence microscopy. Original magnification= 400 X. B. Cells were transiently transfected with 20 pmol of ER β siRNA, incubated 24 h and then fixed and labeled for immunocytochemistry. Upper panel: Control (non-transfected cells). Left: ER β specific green fluorescence in mitochondria and perinuclear compartments of the cells; center: mitochondrial localization of C2C12 cells with the mitochondrial marker MitoTracker Red rendered a red signal; right: merged image of ER β green fluorescence and mitochondria red fluorescence. Lower panel: cells transiently transfected with ER β siRNA; left: weak ER β staining is detected in transfected cells; center: mitochondrial localization with MitoTracker Red; right: merged image. From: Milanesi *et al*., 2009.

regulation of E2 function. Understanding of this network could be of importance to develop therapeutic strategies to overcome pathologies associated with hormonal deregulation. Furthermore, it will help to understand the steroid hormone functions, especially in organs that are not the traditional steroid hormone targets.

CONSENT FOR PUBLICATION

Not applicable.

CONFLICT OF INTEREST

The authors confirm that this chapter contents have no conflict of interest.

ACKNOWLEDGEMENTS

National University of the South Argentina and National Research Council of Argentina (CONICET).

REFERENCES

[1]   Menasce LP, White GR, Harrison CJ, Boyle JM. Localization of the estrogen receptor locus (ESR) to chromosome 6q25.1 by FISH and a simple post-FISH banding technique. Genomics 1993; 17(1): 263-5.
        [http://dx.doi.org/10.1006/geno.1993.1320] [PMID: 8406468]

[2]   Enmark E, Pelto-Huikko M, Grandien K, *et al.* Human estrogen receptor beta-gene structure, chromosomal localization, and expression pattern. J Clin Endocrinol Metab 1997; 82(12): 4258-65.
        [PMID: 9398750]

[3]   Matthews J. Estrogen signaling: a subtle balance between ER alpha and ER beta. Interv 2003; 3: 281.
        [http://dx.doi.org/10.1124/mi.3.5.281]

[4]   Pettersson K, Warner M. Mechanisms of estrogen action. A Physiol Rev 2001; 81: 1535.
        [http://dx.doi.org/10.1152/physrev.2001.81.4.1535] [PMID: 11581496]

[5]   Couse JF, Korach KS. Estrogen receptor null mice: What have we learned and where will they lead us? Endocr Rev 1999; 20(3): 358-417.
        [http://dx.doi.org/10.1210/edrv.20.3.0370] [PMID: 10368776]

[6]   Hammes SR, Levin ER. Extranuclear steroid receptors: Nature and actions. Endocr Rev 2007; 28(7): 726-41.
        [http://dx.doi.org/10.1210/er.2007-0022] [PMID: 17916740]

[7]   Levin ER. Cellular functions of plasma membrane estrogen receptors. Steroids 2002; 67(6): 471-5.
        [http://dx.doi.org/10.1016/S0039-128X(01)00179-9] [PMID: 11960623]

[8]   Figtree GA, McDonald D, Watkins H, Channon KM. Truncated estrogen receptor alpha 46-kDa isoform in human endothelial cells: relationship to acute activation of nitric oxide synthase. Circulation 2003; 107(1): 120-6.
        [http://dx.doi.org/10.1161/01.CIR.0000043805.11780.F5] [PMID: 12515753]

[9]   Driggers PH, Segars JH. Estrogen action and cytoplasmic signaling pathways. Part II: the role of growth factors and phosphorylation in estrogen signaling. Trends Endocrinol Metab 2002; 13(10): 422-7.
        [http://dx.doi.org/10.1016/S1043-2760(02)00634-3] [PMID: 12431838]

[10]   Norfleet AM, Thomas ML, Gametchu B, Watson CS. Estrogen receptor-alpha detected on the plasma membrane of aldehyde-fixed GH3/B6/F10 rat pituitary tumor cells by enzyme-linked immunocytochemistry. Endocrinology 1999; 140(8): 3805-14.
[http://dx.doi.org/10.1210/endo.140.8.6936] [PMID: 10433242]

[11]   Razandi M, Pedram A, Greene GL, Levin ER. Cell membrane and nuclear estrogen receptors (ERs) originate from a single transcript: studies of ERalpha and ERbeta expressed in Chinese hamster ovary cells. Mol Endocrinol 1999; 13(2): 307-19.
[PMID: 9973260]

[12]   Razandi M, Pedram A, Merchenthaler I, Greene GL, Levin ER. Plasma membrane estrogen receptors exist and functions as dimers. Mol Endocrinol 2004; 18(12): 2854-65.
[http://dx.doi.org/10.1210/me.2004-0115] [PMID: 15231873]

[13]   Migliaccio A, Di Domenico M, Castoria G, *et al.* Tyrosine kinase/p21ras/MAP-kinase pathway activation by estradiol-receptor complex in MCF-7 cells EMBO J 1996; 15: 1292.

[14]   Chambliss KL, Simon L, Yuhanna IS, Mineo C, Shaul PW. Dissecting the basis of nongenomic activation of endothelial nitric oxide synthase by estradiol: role of ER αlpha domains with known nuclear functions. Mol Endocrinol 2004; 19: 277.

[15]   Song RX, McPherson RA, Adam L, *et al.* Linkage of rapid estrogen action to MAPK activation by ERalpha-Shc association and Shc pathway activation. Mol Endocrinol 2002; 16(1): 116-27.
[PMID: 11773443]

[16]   Kraus WL, McInerney EM, Katzenellenbogen BS. Ligand-dependent, transcriptionally productive association of the amino- and carboxyl-terminal regions of a steroid hormone nuclear receptor. Proc Natl Acad Sci USA 1995; 92(26): 12314-8.
[http://dx.doi.org/10.1073/pnas.92.26.12314] [PMID: 8618892]

[17]   Acconcia F, Ascenzi P, Fabozzi G, Visca P, Marino M. S-palmitoylation modulates human estrogen receptor-alpha functions. Biochem Biophys Res Commun 2004; 316(3): 878-83.
[http://dx.doi.org/10.1016/j.bbrc.2004.02.129] [PMID: 15033483]

[18]   Acconcia F, Ascenzi P, Bocedi A, *et al.* Palmitoylation-dependent estrogen receptor α membrane localization: regulation by 17β-estradiol. Mol Biol Cell 2005; 16(1): 231-7.
[http://dx.doi.org/10.1091/mbc.e04-07-0547] [PMID: 15496458]

[19]   Li L, Haynes MP. Plasma membrane localization and function of the estrogen receptor -variant (ER46) in human endothelial cells. Proc Natl Acad Sci USA 2003; 100: 4807.

[20]   Wong CW, McNally C, Nickbarg E, Komm BS, Cheskis BJ. Estrogen receptor-interacting protein that modulates its nongenomic activity-crosstalk with Src/Erk phosphorylation cascade. Proc Natl Acad Sci USA 2002; 99(23): 14783-8.
[http://dx.doi.org/10.1073/pnas.192569699] [PMID: 12415108]

[21]   Song RX, Barnes CJ, Zhang Z, Bao Y, Kumar R, Santen RJ. The role of Shc and insulin-like growth factor 1 receptor in mediating the translocation of estrogen receptor alpha to the plasma membrane. Proc Natl Acad Sci USA 2004; 101(7): 2076-81.
[http://dx.doi.org/10.1073/pnas.0308334100] [PMID: 14764897]

[22]   Wyckoff MH, Chambliss KL, Mineo C, *et al.* Plasma membrane estrogen receptors are coupled to endothelial nitric-oxide synthase through Galpha(i). J Biol Chem 2001; 276(29): 27071-6.
[http://dx.doi.org/10.1074/jbc.M100312200] [PMID: 11369763]

[23]   Lu Q, Pallas DC, Surks HK, Baur WE, Mendelsohn ME, Karas RH. Striatin assembles a membrane signaling complex necessary for rapid, nongenomic activation of endothelial NO synthase by estrogen receptor alpha. Proc Natl Acad Sci USA 2004; 101(49): 17126-31.
[http://dx.doi.org/10.1073/pnas.0407492101] [PMID: 15569929]

[24]   Kelly MJ, Qiu J, Wagner EJ, Rønnekleiv OK. Rapid effects of estrogen on G protein-coupled receptor activation of potassium channels in the central nervous system (CNS). J Steroid Biochem Mol Biol

2002; 83(1-5): 187-93.
[http://dx.doi.org/10.1016/S0960-0760(02)00249-2] [PMID: 12650715]

[25] Milanesi L, Russo de Boland A, Boland R. Expression and localization of estrogen receptor a in C2C12 murine muscle cell line J Cell Biochem 2008; 104: 1254-73.
[http://dx.doi.org/10.1002/jcb.21706]

[26] Milanesi L, Vasconsuelo A, Russo de Boland A, Boland R. Expression of Estrogen receptor β in C2C12 cell line and murine skeletal muscle tissue. Steroids 2009; 74: 489-97.
[http://dx.doi.org/10.1016/j.steroids.2009.01.005] [PMID: 19428437]

[27] Levin ER. Integration of the extranuclear and nuclear actions of estrogen. Mol Endocrinol 2005; 19: 1951.
[http://dx.doi.org/10.1210/me.2004-0390]

CHAPTER 3

Estrogens and Androgens Binding Sites in Mitochondria

Lorena M. Milanesi[*]

Instituto de Ciencias Biológicas y Biomédicas de Sur (INBIOSUR)-UNS-CONICET, Bahía Blanca, Argentina

Abstract: Mitochondrial localization of estrogen and androgen receptors and hormone-responsive elements (EREs) was reported in different cell lines and indicated that these steroid hormones have specific actions on the mitochondrial gene expression and functions. Both steroids act through a complex molecular mechanism that involves crosstalk between plasma membrane, mitochondria and nucleus. One of the results of these interactions is mitochondrial protection. Here, we discuss studies that describe the estrogen- and testosterone-dependent actions on the mitochondrial and their implications.

Keywords: 17β− Estradiol (E2), Aging, Mitochondria, Testosterone (T).

INTRODUCTION

Mitochondria is a key regulator of cell survival and death. Indeed, mitochondria manage energy production, free-radical formation, and apoptosis [1, 2].

The existence of ERs and ARs in mitochondria was demonstrated by competitive binding assays in mitochondrial fractions [3 - 5].

The purification of receptor proteins and the accessibility of the corresponding antibodies permitted the application of immunological techniques for the subcellular localization of estrogen and androgen receptors in mitochondria [3 - 10]. Mitochondrial DNA (mtDNA) was also described as one of the major targets for steroid hormones and their receptors [4, 11].

[*] **Corresponding author Lorena M. Milanesi:** Instituto de Ciencias Biológicas y Biomédicas de Sur (INBIOSUR)-UNS-CONICET; Tel: +54 291 4595101x4337; E-mail: milanesi@criba.edu.ar

Andrea Vasconsuelo (Ed.)

E2 Binding Sites in Mitochondria

Presence of mitochondrial 17β–estradiol (E2) binding sites, estrogen receptors and estrogen-response elements in the mtDNA has been demonstrated [3 - 10]. In accordance with this, studies have been done to evaluate E2 functions on this organelle.

The results obtained confirm that mitochondria are estrogen target that affects the mitochondrial morphology [12], electron transport chain [13], mitochondrial permeability transition pore (MPTP) [14] mitochondrial membrane potential [15] and gene expression [13].

As the mitochondria have a central role in apoptosis, it is conceivable that estrogens exert a regulatory action on cellular death. E2 can sustain survival or induce apoptosis of cells depending on their concentration and biological context [1, 16 - 19].

E2 had shown an important antiapoptotic effect mainly at the mitochondrial level, through both ERs (α and β) involving the PI3K/Akt/Bad pathway, Bax, HSP27, MAPKs and various other kinases [4, 12, 14, 15, 20].

E2 also reduces apoptosis in breast cancer cells by bad inactivation [21]; acts as a neuroprotective agent against pathologies such as cardiovascular conditions, in part due to their action on mitochondria [22 - 24].

The mitochondrial electron-transport chain is the main source of reactive oxygen species (ROS) [25].

Chronic ROS exposure generates mtDNA mutations that accumulate during age, but mitochondrial dysfunction seems to be the main contributor to those age-related disorders of the cardiovascular system and the brain [26 - 28].

Estrogens can reduce mitochondrial ROS production [29, 30]. It has been shown that the antioxidant effect due, not only to their chemical phenolic structure, but also to their interaction through ERs that favor nuclear gene expression of antioxidant enzymes which are focused into the mitochondria, such as superoxide dismutase-2 (SOD2) [15, 31, 32].

Thus, this could be part of the mechanism by which estrogens delay the onset of the age-related disorders.

The presence of ERs in mitochondria as it was mentioned before [3] supports the hypothesis that estrogens exert direct and indirect effects on the organelle activity by an intricate signaling network between the plasma membrane, mitochondria and nucleus and the cytoskeleton could play a significant role in the preservation of these interactions [33].

AR has been identified in mitochondria of LNCaP cells, mouse skeletal muscle cell line (C2C12) and in the midpiece of human sperm cells, the piece that contains the mitochondria [5], but less information exists about its mitochondrial localization function.

Testosterone (T) effects on mitochondrial functions have been reported. It has been demonstrated that androgens are capable of depolarizing and oxidizing cardiac mitochondria in a K^+-dependent, ATP-sensitive, and AR–independent mode.

In C2C12 cell line, H_2O_2 induces the mitochondrial permeability transition pore opening (mPTP) and p53 activation, with a maximum response after 1-2 hours of H_2O_2 exposure. Testosterone treatment, prior to H_2O_2, reduces not only p53 phosphorylation but also the mPTP opening [34].

Mitochondria integrate a great number of signal transduction pathways and enclose various pro-apoptotic proteins, representing a control point of apoptosis that is controlled by members of the Bcl-2 family (Fig. **1**). Then, if mitochondria contain estrogen and androgen binding sites, the steroid hormones could modulate programmed cell death.

CONCLUDING REMARKS

It is clear that these steroid hormones apply their activities to the mitochondria and play an important role in age-related processes. Both steroids affect the mitochondrial function directly (through ERs, ARs and HREs that localized in the organelle and regulate gene expression of mitochondrial proteins) and/or indirectly (by the regulation of nuclear transcription factors that affect mitochondrial DNA-encoded proteins). Also the hormones switch various mitochondrial functions such as ROS generation and apoptosis signaling. The cytoskeleton could play a crucial role in the articulation of all these mechanisms (Fig. **1**).

A full understanding of the molecular mechanism started by both hormones at the mitochondrial level brings capacities for the future treatment of age-dependent diseases related to the deregulation of sexual hormones.

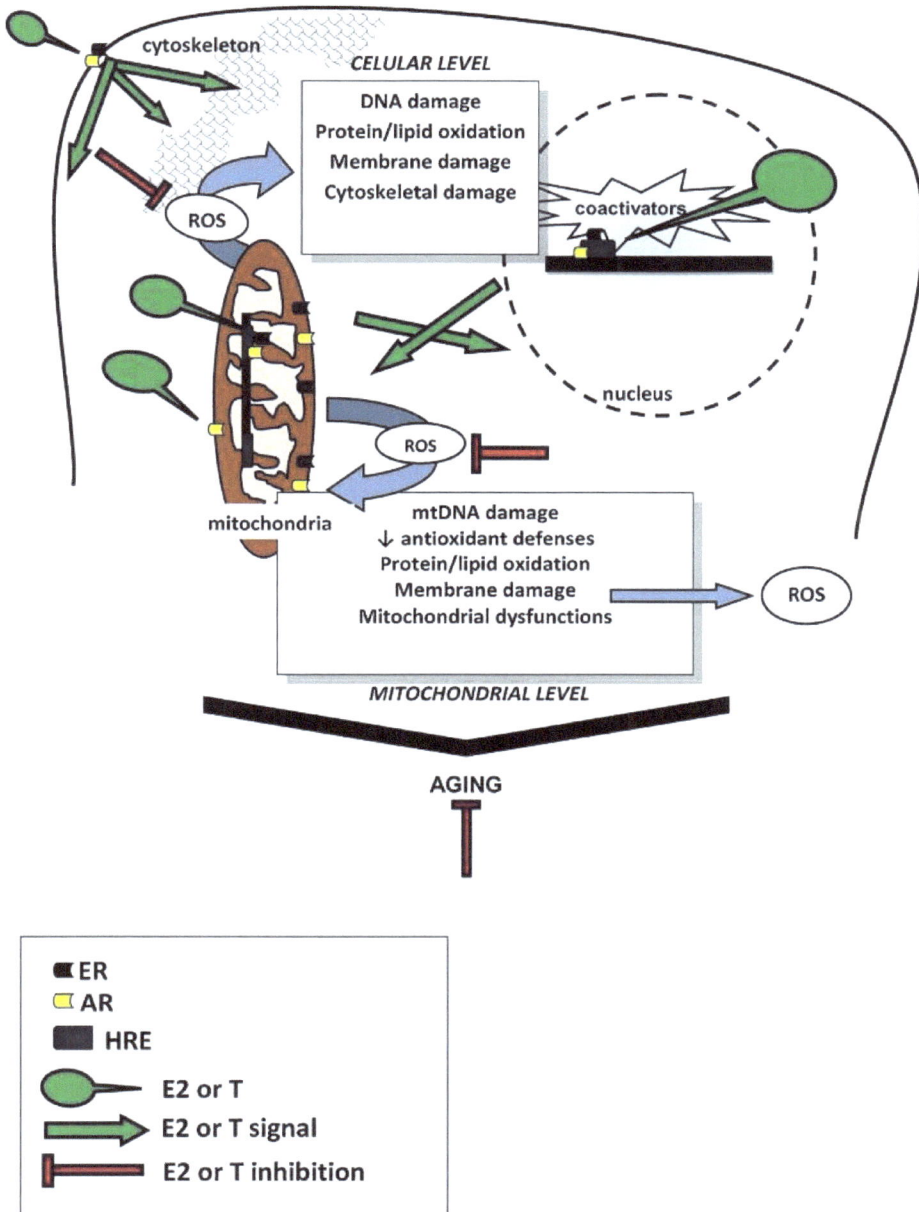

Fig. (1). Effect of E2 and T on mitochondria. Mitochondria are a source of reactive oxygen species (ROS). ROS bring a decline in mitochondrial function that results in more oxidative stress. ROS production disturbs the hole cell, leading to mtDNA and nuclear DNA damage, decrease of antioxidant defenses, cytoskeletal injury, protein/lipid oxidation as well as membrane damage. E2 and T could act through their different subcellular locations (membrane, mitochondria, cytosol and nucleolus) against their specific HREs; establishing an intricate signaling network which results in a protective action [Extracted from 35].

CONSENT FOR PUBLICATION

Not applicable.

CONFLICT OF INTEREST

The authors confirm that this chapter contents have no conflict of interest.

ACKNOWLEDGEMENTS

National University of the South Argentina and National Research Council of Argentina (CONICET).

REFERENCES

[1]   Green DR, Reed JC. Mitochondria and apoptosis. Science 1998; 281(5381): 1309-12.
        [http://dx.doi.org/10.1126/science.281.5381.1309] [PMID: 9721092]

[2]   Dimmer KS, Scorrano L. (De) constructing mitochondria: what for? Physiology 2006; 21: 233-41.

[3]   Milanesi L, Russo de Boland A, Boland R. Expression and localization of estrogen receptor a in C2C12 murine muscle cell line. J Cell Biochem 2008; 104: 1254-73. ISSN 0730-2312

[4]   Milanesi L, Vasconsuelo A, Russo de Boland A, Boland R. Expression of Estrogen receptor β in C2C12 cell line and murine skeletal muscle tissue. Steroids 2009; 74: 489-97.

[5]   Non-classical localization of androgen binding proteins in the C2C12 skeletal muscle cell line Archives of Biochemistry and Biophysics 2013; 13-22.

[6]   Monje P, Boland R. Characterization of membrane estrogen binding proteins from rabbit uterus. Mol Cell Endocrinol 1999; 147(1-2): 75-84.
        [http://dx.doi.org/10.1016/S0303-7207(98)00212-3] [PMID: 10195694]

[7]   Monje P, Boland R. Subcellular distribution of native estrogen receptor alpha and beta isoforms in rabbit uterus and ovary. J Cell Biochem 2001; 82(3): 467-79.
        [http://dx.doi.org/10.1002/jcb.1182] [PMID: 11500923]

[8]   Chen JQ, Eshete M, Alworth WL, Yager JD. Binding of MCF-7 cell mitochondrial proteins and recombinant human estrogen receptors alpha and beta to human mitochondrial DNA estrogen response elements. J Cell Biochem 2004; 93(2): 358-73.
        [http://dx.doi.org/10.1002/jcb.20178] [PMID: 15368362]

[9]   Chen JQ, Delannoy M, Cooke C, Yager JD. Mitochondrial localization of ERalpha and ERbeta in human MCF7 cells. Am J Physiol Endocrinol Metab 2004; 286(6): E1011-22.
        [http://dx.doi.org/10.1152/ajpendo.00508.2003] [PMID: 14736707]

[10]   Yang S H, Liu R, Perez E J, *et al.* Mitochondrial localization of estrogen receptor beta. Proc Natl Acad Sci USA 2004; 101: 4130-5.

[11]   Demonacos CV, Karayanni N, Hatzoglou E, Tsiriyiotis C, Spandidos DA, Sekeris CE. Mitochondrial genes as sites of primary action of steroid hormones. Steroids 1996; 61(4): 226-32.
        [http://dx.doi.org/10.1016/0039-128X(96)00019-0] [PMID: 8733006]

[12]   Vasconsuelo A, Milanesi LM, Boland RL. 17β-Estradiol abrogates apoptosis in murine skeletal muscle cells through estrogen receptors: role of the phosphatidylinositol 3-kinase/Akt pathway J Endocrinol 2008; 196: 385-97.

[13]   Chen JR, Plotkin LI, Aguirre JI, *et al.* Transient versus sustained phosphorylation and nuclear accumulation of ERKs underlie anti-versus pro-apoptotic effects of estrogens. J Biol Chem 2005;

280(6): 4632-8.
[http://dx.doi.org/10.1074/jbc.M411530200] [PMID: 15557324]

[14]   La Colla A, Vasconsuelo A, Boland R. Estradiol exerts antiapoptotic effects in skeletal myoblasts via mitochondrial PTP and MnSOD. J Endocrinol 2013; 216: 331-41.

[15]   La Colla A, Boland R, Vasconsuelo A. 17β-estradiol abrogates apoptosis inhibiting PKCδ, JNK, and p66Shc activation in C2C12 Cells. J Cell Biochem 2015; 116(7): 1454-65.
[http://dx.doi.org/10.1002/jcb.25107] [PMID: 25649128]

[16]   Okasha S, Ryu S, Do Y, McKallip R,J, Nagarkatti M, Nagarkatti P,S. Evidence for estradiol-induced apoptosis and dysregulated T cell maturation in the thymus. Toxicol 2001; 163: 49-62.
[http://dx.doi.org/10.1016/S0300-483X(01)00374-2]

[17]   Florian M, Magder S. Estrogen decreases TNF-alpha and oxidized LDL induced apoptosis in endothelial cells. Steroids 2008; 73: 47-58.

[18]   Seli E, Guzeloglu-Kayisli O, Kayisli UA, Kizilay G, Arici A. Estrogen increases apoptosis in the arterial wall in a murine atherosclerosis model. Fertil Steril 2007; 88(4) (Suppl.): 1190-6.
[http://dx.doi.org/10.1016/j.fertnstert.2007.01.132] [PMID: 17498707]

[19]   Choi KC, Kang SK, Tai CJ, Auersperg N, Leung PC. Estradiol up-regulates antiapoptotic Bcl-2 messenger ribonucleic acid and protein in tumorigenic ovarian surface epithelium cells. Endocrinology 2001; 142: 2351-60.

[20]   Vasconsuelo A, Milanesi L, Boland R. Participation of HSP27 in the antiapoptotic action of 17beta-estradiol in skeletal muscle cells. Cell Stress Chaperones 2010; 15: 183-92.

[21]   Fernando RI, Wimalasena J. Estradiol abrogates apoptosis in breast cancer cells through inactivation of BAD: Ras-dependent nongenomic pathways requiring signaling through ERK and Akt. Mol Biol Cell 2004; 15(7): 3266-84.
[http://dx.doi.org/10.1091/mbc.e03-11-0823] [PMID: 15121878]

[22]   Bishop J, Simpkins JW. Estradiol treatment increases viability of glioma and neuroblastoma cells *in vitro.* Mol Cell Neurosci 1994; 5(4): 303-8.
[http://dx.doi.org/10.1006/mcne.1994.1036] [PMID: 7804599]

[23]   Behl C, Widmann M, Trapp T, Holsboer F. 17-beta estradiol protects neurons from oxidative stress-induced cell death *in vitro.* Biochem Biophys Res Commun 1995; 216(2): 473-82.
[http://dx.doi.org/10.1006/bbrc.1995.2647] [PMID: 7488136]

[24]   Zhai P, Eurell TE, Cotthaus R, Jeffery EH, Bahr JM, Gross DR. Effect of estrogen on global myocardial ischemia-reperfusion injury in female rats. Am J Physiol Heart Circ Physiol 2000; 279(6): H2766-75.
[http://dx.doi.org/10.1152/ajpheart.2000.279.6.H2766] [PMID: 11087231]

[25]   Lesnefsky E, Moghaddas S, Tandler B, Kerner J. Mitochondrial dysfunction in cardiac disease: ischemia--reperfusion, aging, and heart failure. J Mol Cell Cardiol 2001; 33: 1065.
[http://dx.doi.org/10.1006/jmcc.2001.1378]

[26]   Wallace DC. A mitochondrial paradigm of metabolic and degenerative diseases, aging, and cancer: a dawn for evolutionary medicine. Annu Rev Genet 2005; 39: 359-407.
[http://dx.doi.org/10.1146/annurev.genet.39.110304.095751] [PMID: 16285865]

[27]   Madamanchi NR, Runge MS. Mitochondrial dysfunction in atherosclerosis. Circ Res 2007; 100(4): 460-73.
[http://dx.doi.org/10.1161/01.RES.0000258450.44413.96] [PMID: 17332437]

[28]   Abbott NJ, Rönnbäck L, Hansson E. Astrocyte-endothelial interactions at the blood-brain barrier. Nat Rev Neurosci 2006; 7(1): 41-53.
[http://dx.doi.org/10.1038/nrn1824] [PMID: 16371949]

[29]   Pedram A, Razandi M, Wallace DC, Levin ER. Functional estrogen receptors in the mitochondria of

breast cancer cells. Mol Biol Cell 2006; 17(5): 2125-37.
[http://dx.doi.org/10.1091/mbc.e05-11-1013] [PMID: 16495339]

[30]   Razmara A, Duckles SP, Krause DN, Procaccio V. Estrogen suppresses brain mitochondrial oxidative stress in female and male rats. Brain Res 2007; 1176: 71-81.
[http://dx.doi.org/10.1016/j.brainres.2007.08.036] [PMID: 17889838]

[31]   Borras C, Sastre J, Garcia-Sala D, Lloret A, Pallardo F, Vina J. Direct antioxidant and protective effect of estradiol on isolated mitochondria. Free Radic Biol Med 2003; 34: 546.
[PMID: 12614843]

[32]   Vina J, Borras C, Gomez-Cabrera MC, Orr WC. Part of the series: from dietary antioxidants to regulators in cellular signalling and gene expression. Role of reactive oxygen species and (phyto)oestrogens in the modulation of adaptive response to stress. Free Radic Res 2006; 40(2): 111-9.
[http://dx.doi.org/10.1080/10715760500405778] [PMID: 16390819]

[33]   Carré M, André N, Carles G, *et al.* Tubulin is an inherent component of mitochondrial membranes that interacts with the voltage-dependent anion channel. J Biol Chem 2002; 277(37): 33664-9.
[http://dx.doi.org/10.1074/jbc.M203834200] [PMID: 12087096]

[34]   Pronsato L, Milanesi L. Effect of testosterone on the regulation of p53 and p66Shc during oxidative stress damage in C2C12 cells. Steroids 2016; 106: 41-54.
[http://dx.doi.org/10.1016/j.steroids.2015.12.007] [PMID: 26703444]

[35]   Vasconsuelo A, Milanesi L, Boland R. Actions of 17β-estradiol and testosterone in the mitochondria and their implications in aging. Ageing Res Rev 2013; 12(4): 907-17.
[http://dx.doi.org/10.1016/j.arr.2013.09.001] [PMID: 24041489]

CHAPTER 4

General Concepts of Skeletal Muscle and Apoptosis: Molecular Mechanisms and Regulation by Sex Steroids

Andrea Vasconsuelo[*] and **Lucia Pronsato**[*]

Instituto de Ciencias Biológicas y Biomédicas del Sur (INBIOSUR), Universidad Nacional del Sur- CONICET, Bahía Blanca, Argentina

Abstract: Apoptosis is a physiologic process that take place during development and in the progression of specific diseases. In skeletal muscle, this process is poorly explored. Skeletal muscle represents an exceptional tissue regards apoptosis, by its multinucleated structure and its variable mitochondrial content. On the other hand, apoptosis of skeletal muscle tissue could have a wide spectrum of effects on organism since it is now well established that skeletal muscle not only generates force and movement; other functions are associated to this tissue. Skeletal muscle contributes to basal energy metabolism, in the storing for substrates such as amino acids and carbohydrates, in the keep of core temperature and blood glucose levels, and in the use of oxygen and energy during movement. Also, skeletal muscle acts as an endocrine organ, thus could be regulated by owning or no own hormones. This chapter summarizes the generalities of skeletal muscle at the molecular/structural and functional level and basic concepts of apoptosis, for a better understanding of the following chapters of this ebook. Which it focuses specifically, at a molecular level involving genomic regulation, on the actions of 17β-Estradiol and Testosterone in the homeo- stasis of skeletal muscle tissue in physiological and pathological conditions, converging in the relationship of muscle apoptosis and sexual hormones.

Keywords: Apoptosis Pathways, Bcl-2 Family, Caspases, Muscle Structure, p53-Regulated Genes, Skeletal Muscle.

INTRODUCTION

The kinds of muscle tissue are: skeletal, smooth and cardiac; being myocytes the cellular unit of all types of muscle. Myocytes contain specialized proteins that use chemical energy to generate mechanical force as cellular contraction, the elementary action of all the functions exercised by the muscle [1].

[*] **Corresponding authors Andrea A. Vasconsuelo and Lucía Pronsato:** Instituto de Ciencias Biológicas y Biomédicas del Sur (INBIOSUR), Universidad Nacional del Sur- CONICET, Bahía Blanca, Argentina; Tel: +54 291 4595101x4337; E-mails: avascon@criba.edu.ar; lpronsato@criba.edu.ar

Concepts of Skeletal Muscle is mainly distributed in the extremities. In humans, it is the largest organ in the body and consists of fibers, nervous cells, vasculature nets, and connective tissue. The myofibers (multinucleated contractile muscle cells) work like structural/functional constituents of skeletal muscle. Bundles of myofibers named as fascicles build the muscle tissue. The envelope surrounding the muscle fiber is the sarcolemma, and this membrane encloses the sarcoplasm, containing the thin and thick filaments of actin and myosin respectively [Review in 2]. The muscular actin is a filamentous protein and myosin has a filamentous tail and one or two heads in which exist binding sites for actin [1]. Myosin heavy chain (MyHC) is the key contractile protein of skeletal muscle; it is fundamental for the efficiency of muscle contraction. Muscle fibers differ in types of MyHC, the fast or slow isoforms. Also differ in the class of metabolism (oxidative or glycolytic). Hormones regulate the selection of myosin gene expression. Moreover, myofibers can be clustered into two groups: one is the slow-contracting/fatigue-resistant and the other is the fast-contracting/fatigu--susceptible, according to their physio- logical properties [reviewed in 3].

The expression of myosin gene is also controlled by intronic microRNAs, such as miR-208b and miR-499, present in MyHC genes. These microRNAs can play important regulatory roles by pointing mRNAs for destruction or translational repression. MicroRNAs encompass one of the more rich types of gene regulatory particles and likely impact on the yield of many protein-coding genes [4]. Thus, genes encoding myosin also affect muscle role through a network of intronic miRNAs that regulate muscle gene expression and function. The importance, associations of microRNA regulation with sex steroids have been strongly suggested [reviewed 5], implicating a key role of sex hormones on skeletal muscle homeostasis.

The organization of actin and myosin fibers provides skeletal muscle its striated form [review in 2. Briefly, sarcomeres are the contractile units of a muscle and are composed of the two types of filaments: thick and thin. There are three key properties of sarcomere which are critical to its function: (1) its capability to shorten fast and efficiently, (2) its capacity to turn on and off very fast, and (3) its accuracy self-assembly and structural regularity. Each sarcomere is delimited by a dark stripe called a Z disc. Alpha-actinin (Mr = 97 kDa) is one of the major components of the Z disc. It is a member of the dystrophin superfamily [Review in 6]. The thin filaments are attached at one end to a Z disc and extend toward the center of the sarcomere. The thick filaments, lie at the midpoint of the sarcomere and overlap the thin filaments. Other proteins, together with Z disc, are involved in keeping the architecture of the sarcomere [Review in 6].

At the physiological level, skeletal muscle has one key task; namely, to transform chemical energy taken from the food into mechanical energy; this process is activated by the nervous system [1]. The physiology of skeletal muscle is the basis for understanding animal movement. The elementary mechanism of force and movement generation was suggested in 1954. That principle is recognized as the "sliding filament theory" and remains to be, with some adjustments, the accepted elucidation of muscle contraction [7]. Broadly, the theory proposes that the thick and thin filaments slide with respect to one another, using ATP as a source of energy and responding to the signals from motor neurons. The contraction of a whole muscle fiber results from the simultaneous contraction of all of its sarcomeres. Muscular contraction involves origination and propagation of a membrane action potential, transduction of the electrical energy into an intracellular chemical signal that, in sequence, elicits filaments contact. The stimulus for physiological skeletal muscle action born in a nerve impulse [8]. Briefly, motor neuron influences several muscle cells in an assemblage identified as a motor unit. The neuromuscular junction is the point where motor neurons contact myofiber. When a motor neuron acquires a signal from the central nervous system, it excites all of the fibers in its motor unit at the same time. The neurons release neurotransmitters that link to sarcolemma in the motor end plate. This specific zone in the sarcolemma contains numerous ion channels that open in response to neurotransmitters and allow positive ions to enter the muscle fiber, forming an electrochemical gradient inside of the cell, which extents all through the plasmatic membrane. The cations reach the sarcoplasmic reticulum, then Ca^{2+} ions are released into the myofibrils and bind to troponin causing a conformational change, which causes the troponin molecule travel near of tropomyosin. Tropomyosin has moved away from myosin binding sites on actin molecules, allowing the interaction actin-myosin. ATP molecules power myosin proteins to bend and pull on actin molecules. As a result, the actin filaments are pulled closer to the middle of a sarcomere. As the thin filaments are pulled together, the sarcomere reduces and contracts. Myofibrils are made up of numerous sar- comeres, so that when all of the sarcomeres contracts, the myofiber shortens. The body can regulate the strength of each muscle determining the number of motor units that activates for a given role or activity. Muscles remain contraction while they are stimulated by a neurotransmitter. Once a neuron ends the discharge of the neuromessenger, the contraction backs itself. The ions Ca^{2+} returns to the intracellular reservoirs mainly to the sarcoplasmic reticulum; troponin and tropomyosin return to their inactive state and actin and myosin are disallowed to binding. Sarcomeres return to their elongated resting state once the force of myosin pulling on actin has finished.

Skeletal muscle is a form of striated muscle that provides vertebrates with a locomotive ability that impacts on activity, allows for participation in social life and work-related habits contributing to functional independence. Indeed the main function of skeletal muscle generates force and power, maintains pose, and produces movement. It is for this reason that in the senescence the progressive loss of muscle performance [9] affects the routine movements and independence in the elderly [10, 11]. However, not only movements and independence are affected when the muscle loses its functional capacity since other roles have been associated with this tissue. This organ comprises nearly 40% of entire body weight, contains 50–75% of all body proteins, and accounts for 30–50% of whole-body protein turnover. Accordingly, skeletal muscle is involved in the maintenance to basal energy metabolism, in the storage for important bio-molecules such as protein precursors and carbohydrates, in the regulation of corporal temperature, and in the utilization of the majority of oxygen and energy engaged during physical activity. In addition, associated with the role of storage of amino acids, these releases from this tissue contributes to the keep of glycemia in conditions of severe deficiency in caloric energy intake. Therefore, metabolic flexibility and insulin sensitivity of skeletal muscle contribute greatly to glucose homeostasis and whole body metabolism [12]. Certainly, this tissue is the major site of substrate metabolism. Moreover, evidence indicates another no classical role for skeletal muscle as an endocrine organ [Review in 13]. In general, the hormones produced by skeletal muscle tissue are termed "myokines" and this exerts endocrine effect [14]. "Myokines" are proteins or peptides secreted by muscle tissue to control the metabolic process in muscle and other tissues. Besides the production of myokines, it has been demonstrated that under certain circumstances skeletal muscle is able to synthesize biologically active steroid male and female hormones [15]. It is known that androgens and estrogens are mainly made and secreted by the gonads. However, has been shown that testosterone, estradiol, and 5α-dihydrotestosterone also are locally produced in skeletal muscle from dehydroepiandrosterone in several experimental systems [15, 16]. Thus, skeletal muscle could be classified as an endocrine organ. In addition, muscle is the target of the own sex steroids [Review in 17]. Indeed, has been found in skeletal muscle receptors for both types of steroids androgen and estrogen [18, 19]. It's been known for a long time the relevance of estrogen and testosterone for the health of skeletal muscle [Review in 20, 21]. Although muscle mass depends on the equilibrium of protein synthesis and turnover, process which are subtle to factors such as diet, physical activity, injury or disease, the loss of skeletal muscle mass and functionality in mature adults, is associated with sex hormones decline in this stage of life [Review in 22, 23]. The normal decline in voluntary strength and/or performance of this tissue in the elderly has important health-related consequences; leading to reduced mobility, potentially leading to functional incapacity and institutionalization [24]. In the present text, it will deepen in the hormonal effects on skeletal muscle.

It is evident that skeletal muscle plays various functions in the organism; is responsible for all the voluntary movements, being essential for optimal physical performance and is sensitive to various factors. Accordingly, it can be affected by a variety of pathologies.

The aim of the present eBook is to describe at the molecular level the actions of 17β-Estradiol (E2) and Testosterone (T) in the homeostasis of skeletal muscle tissue in physiological and pathological conditions, focusing on the relationship of muscle apoptosis and sexual hormones.

Basic Concepts of Apoptosis: Molecular Machinery

Apoptosis is a word taken from ancient Greek which means the falling of leaves. In 1972 the term apoptosis is proposed for a little-known process of controlled cell death, which appears as a paired but opposed process to mitosis in the control of cell populations [25]. This process was originally defined by Kerr, Wyllie, and Currie as morphological changes at the nuclear and cytoplasm level. Apoptotic cells exhibit characteristic alterations at nuclear level, with aggregation/condensation of the chromatin round the nuclear envelope followed by disintegration of this membrane and buildup of bodies of aggregated chromatin in the cytoplasm. Beside with these nuclear modifications, the cytoplasm of dying cells contracts and cell size declines due to the loss of cellular content through the injured plasma lemma. Also, the cells fail to contact with the extracellular matrix and disconnect from adjacent cells [25]. The apoptotic course finishes with cellular breakdown into fragments that are phagocyted. Then, the term apoptosis was used to describe those morphologically dissimilar features, as opposed to accidental, cell death; the latter termed necrosis [26]. Investigations have shown that both processes differ in their mechanisms of induction and execution. Necrosis is prompted by toxic chemical, biological or physical events and traditionally was considered an uncontrolled process. Briefly, in necrosis the loss of plasma membrane integrity occurs, with organelle and cell swelling, causing cell lysis; the outflow of intracellular components can injury adjacent or nearby cells and induces an inflammatory response [27]. By contrast, apoptosis is the regulated destruction of a cell and needs the synchronization of gene-directed and energy-dependent biological processes and this event is triggered by specific signals [review in 26].

Apoptosis is a necessary process: during development, immune reactions and aging, to the removal of excess of potentially dangerous cells, tumor cells and/or virus-infected cell [Review in 28].

The basic apoptotic enzymatic apparatus involves a cysteine dependent aspartate-specific protease family, called each member "Caspase" (cysteine-dependent aspartate-directed proteases) [29].

General Pathways

For a better interpretation of the apoptotic process, we can consider three stages, which will vary in different aspects and in mediators involved in each stage dependent on the cell type and/or species considered. Generally, the tumor suppressor gene p53 participates in the first stage and intracellular or extracellular signals activate specific mechanisms that will decide whether a cell should survive or die [Review in 28]. In this context, the extrinsic (death receptor) and intrinsic (mitochondrial) pathways represent the two main more studied apoptotic mechanisms. The second stage involves the action of B-cell lymphoma-2 family of proteins (Bcl-2), a group of apoptotic regulators, and the protein-cleaving enzymes called caspases (cysteine-dependent aspartate-directed proteases), the center of the apoptotic machinery. They are manager of proteolytic cleavage of an extensive range of cellular targets [30]. Caspases exist as zymogen, an inactive state of the apoptotic enzyme, in order to protect the cell from aberrant proteolysis. Thus, procaspases must undergo an activation event with the purpose of continue the apoptotic process. However, news evidence ascribes to caspases a myriad of non-cell death functions [Review in 31]. In the third stage, the cell death process reaches a point of no return, with DNA fragmentation, proteolysis of proteins, and typical morphological alterations as membrane blebbing, mitochondrial and nuclear pyknosis, mitochondrial redistribution and cytoskeletal disorganization among others [Review in 28].

The extrinsic pathway is initiated by the activation, through interactions with specific ligands, of a subgroup of the tumor necrosis factor receptor (TNFR) family also called death receptors [32]. These receptors present a cytoplasmic region known as the "death domain", which transmit the death signals from the exterior cellular to the intracellular signaling pathways. The receptor-ligand interactions induce grouping of receptor and the recruitment with oligomerization of adaptor proteins for instance FADD to the Fas receptor. The caspase-8 zymogen, procaspase-8, links to the FADD by dimerization of their death effector region establishing the death-inducing signaling complex (DISC). Indeed, stimulation of these receptors conduce to the recruitment and activation of initiator caspases 8-10. Initiator caspases modify and activate effector procaspase, classically procaspase 3. The activated caspase 3 cleaves an amount of death targets that lead to typical hallmarks of apoptosis, above mentioned [Review in 33; Review in 28].

Unlike the extrinsic apoptotic *via*, the mitochondrial pathway can be triggered by an array of physical, chemical and pathophysiological signals. The best stimuli studied include DNA damage, metabolic stress, cellular stress, cytokine deficiency, UV radiation, heat, hypoxia, and chemotherapeutic drugs, all of them converge at the mitochondria [Review in 28]. Under those stimuli, mitochondria undergo dramatic changes in their structure and function, taking a crucial role in the intrinsic pathway [Review in 34]. This central action of mitochondria is ruled by the Bcl-2 proteins (Fig. **1**), which are distributed into three characteristic subgroups. The first is the anti-apoptotic members, Bcl-2, Bcl-w, Bcl-xL, Mcl-1, and A1. The second is two pro-apoptotic members Bak and Bax. The pro-apoptotic cluster can be more divided as either multi-BH-domain proteins, including Bax and Bak, or as BH3-only proteins, for example Bid, Bad, Bim, Puma, and Noxa. In response to pro-apoptotic events, a third group the so-called 'BH3-only' proteins, turn into activated state and overwhelm the antiapoptotic proteins to allow Bak and Bax activation. Actually, somehow all members of the Bcl-2 family control the permeability of the outer mitochondrial membrane. Active Bak and Bax induce outer-membrane permeabilization in mitochondria, that is done by the creation of a pore or conduct in the mitochondrial outer membrane and this triggers the efflux of cytochrome c. So, the cautiously controlled equilibrium among pro-apoptotic and anti-apoptotic members of the Bcl-2 family define the endurance of the cell [35].

Cytochrome c is an apoptogenic element, located in mitochondrial intermembrane space [Review in 34]. So, a disproportion between apoptotic and antiapoptotic Bcl-2 proteins, induce cytochrome c release which triggers the establishment of the apoptosome complex by recruiting the adaptor element Apaf-1 and assisting its Interaction with dATP. This induces a conformational transformation in Apaf-1 that exposes its caspase-activation and recruitment region, all these changes allow the participation in the apoptosome of procaspase-9, the zymogen of caspase-9 [36, 37]. The first zymogen to be activated in the mitochondrial pathway is procaspase-9 and is accordingly nominated the initiator caspase of this *via*. Procaspase-9 is transformed to caspase-9 through dimerization and is successively poised to activate downstream effector caspases, by cleavage of the respective zymogens. The effector caspases of the mitochondrial cascade are caspases-3 and -7 [38]. These apoptotic enzymes are required for the cleavage of the important cellular proteins, such as cytoskeletal proteins, that leads to the distinctive morphological alterations detected in apoptotic cells. As was mentioned, one of the hallmarks of programmed cell death is the chromosomal DNA degradation. Also, the caspases play an significant role in this process by activating DNases, inhibiting DNA repair systems and breaking down nuclear proteins.

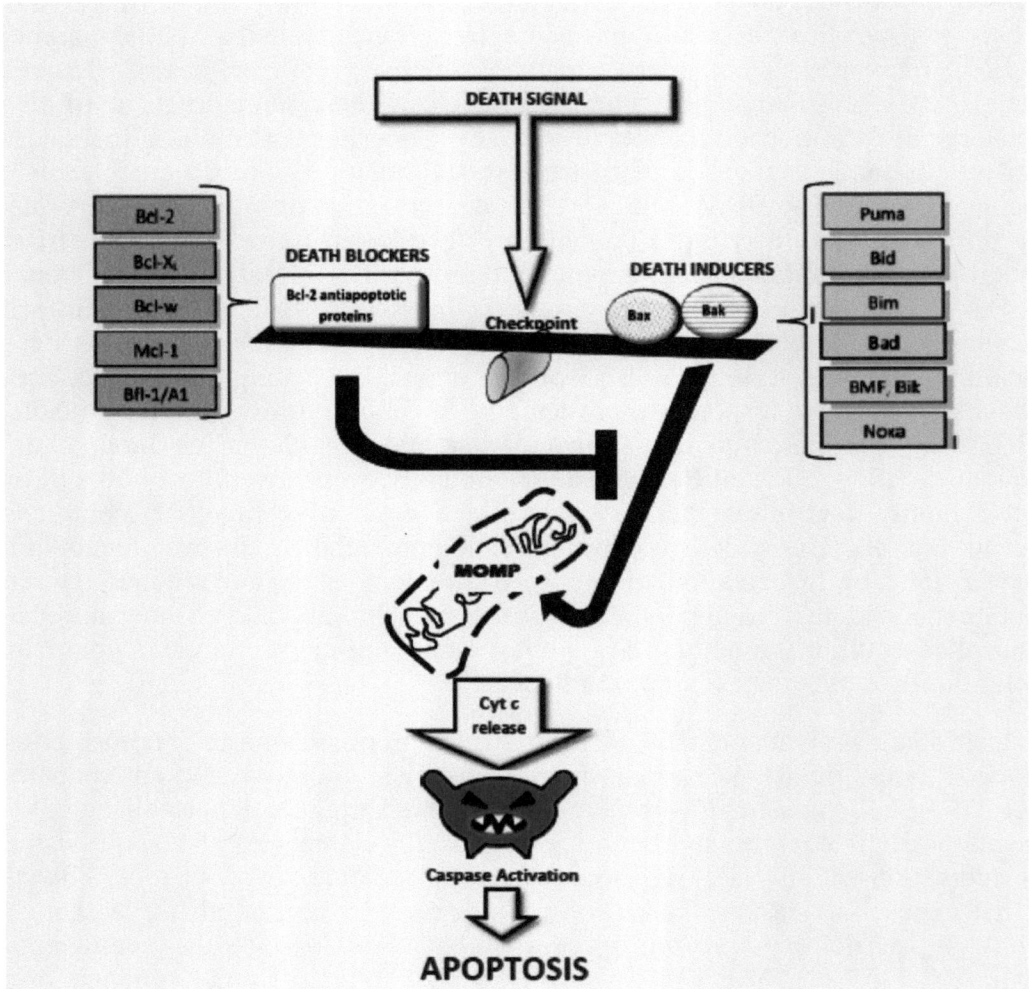

Fig. (1). Control of apoptosis at the mitochondrial level by the Bcl-2 proteins.

The blockers of apoptosis of Bcl-2 family (Death Blockers) physically interact and isolate Bak or Bax. Upon death stimulus, Bak or Bax are released from that interaction. In this point, the BH3-only proteins such as Bid, Bad, Puma, Bim in the others play a key role. Then Bax and Bak oligomerize causing in mitochondrial permeabilization (MOMP) and in consequence the release of apoptotic elements such as cytochrome c. Cytochrome c induce caspase activation causing in apoptosis.

Apoptosis in Skeletal Muscle

As it was mentioned apoptosis is a explicit fashion of cellular death, which ensures the elimination of injured, pathologic or superfluous cells during development and in adulthood; playing thus, a key role in homeostasis. Respect of tissue kinetics may be considered a useful tool to balances the effect of cell proliferation. Although in very proliferative tissues apoptosis keep a constant amount of cells, in postmitotic tissues such as skeletal muscle the role of programmed cell death is less clear. Apoptosis is involved in healthy muscle development since apoptosis and differentiation share common signaling pathways in muscle cells; for example, caspase activity plays a central role in both apoptosis and muscle differentiation. However, if apoptosis increases at inappropriate times or levels, the cellular death process becomes harmful to skeletal muscle [39].

In the previous section, we described characteristics of the skeletal muscle that determines that the process of apoptosis is singular in this tissue. It has been explained the crucial role of cellular nuclei in apoptosis, but in what way is apoptosis coordinated by a multinucleated cell, such as the muscle cell? Indeed, the trigger of the apoptosis results in the elimination of individual myonuclei and the relative portion of sarcoplasm without the dismantling of the whole fiber, a procedure identified as myonuclear apoptosis [40]. To add more complexity to apoptotic mechanisms within skeletal muscle tissue, we must consider that skeletal muscle cell contains a variable number of mitochondria, the other key organelle involved in apoptosis. The quantity of mitochondria in skeletal muscle is affected by the fiber type and the level of training [41]. Moreover, these mitochondria vary morphologically and biochemically. These discrepancies result in diverse susceptibility towards apoptotic stimuli and may then be differentially involved in the origin of sarcopenia. In addition to the factors described before, involved in the control of apoptosis in skeletal muscle, we must consider that muscle cells express receptors for estrogens and androgens in mitochondria [42 - 44]. This discovery led to the demonstration that mitochondrial functions were regulated in some way by estrogens and androgens. In consequence, given the central role of this organelle in apoptosis, hormonal variations would be reflected in the apoptotic processes of the muscle. This fact is in agreement with studies showing that normal or pathologic alterations in the levels of circulating estradiol or testosterone have consequences for muscle homeostasis in both men and women [45]. In the agreement, it is amply recognized that skeletal muscle is a target tissue for male sex hormone. This hormone by its actions on muscle and fat tissue is a crucial determinant of body structure in male mammals, comprising human. Testosterone supplementation growths muscle mass in healthy young and old men, healthy hypogonadal males, as well as other situations with low levels of this hormone [46]. It has also been demonstrate that Testosterone treatment prompted escalation in muscle size is linked with hypertrophy of skeletal muscle fibers and substantial rises in myonuclear and myoblast cells populations [11, 47, 48]. Existing data further suggests that exogenous male hormone administration conduces to recovery from hind limb paralysis after sciatic nerve damage in the rat [49], and wholly avoids the castration-activated apoptosis in muscle cells of the rat levator ani muscle [50]. There is also accumulating evidence that demonstrates that skeletal muscle tissue is responsive to estrogens; where as well as testosterone, it exerts a protective effect against apoptosis [43, 51 - 55]. Really, the arrangement of the effects of the upregulation of antiapoptotic molecules, down regulation of proapoptotic ones and the results of their connections can underwrite to the positive role of these hormones. Thus, the role of sex steroids on apoptotic skeletal muscle, depends also on the regulation of transcriptional functions, that stimulate or upset cellular survival.

Regulation of p53 Activation by Sex Steroids during Apoptosis of Skeletal Muscle

The tumor suppressor agent p53, commonly recognized as 'the guardian of the genome', has a vital function as a damage-control mechanism. The p53 path is encouraged by diverse stimulatory signals that threaten to disturb the integrity of the genome or the appropriate cell proliferation. It detects a variety of cellular stresses and integrates and translates these signals in the most adequate reaction (cell cycle arrest, DNA reparation, senescence, apoptosis) in order to avoid the cell injury extent [56, 57]. P53 acts fundamentally as a transcription factor, transcribing a set of a varied range of genes to achieve its biological reactions [57], in response to many types of cellular stress, including destruction of DNA, energetic withdrawal, oncogenic imbalance, hypoxia and oxidative stress [58]. Once activated, p53 triggers a heterogeneous reaction in a cell-, tissue- and (type and intensity) stress-dependent fashion. The exact amalgamation of the signals received, or the changes that they can induce on the protein, by post-translational variations, directions p53 behavior in any given condition. Consequently, the specific set of circumstances dictates whether the p53 response is dependent or independent of transcription, the timing and array of gene expression transformed, and so manages the destiny of the cell [56]. The stimulation of some sets of target genes is one of the tools by which p53 join in stress signals into a cell reaction. The genetic information implicated in the cell-cycle arrest are the former to have high expression levels, being the apoptotic genes expressed at intermediate and late phases [59]. Low levels of p53 stimulate genes with high-affinity promoters that tend to be linked to cell-cycle arrest, and great levels of p53 activate low-affinity promoters that incline to be involved in the apoptosis [60]. One of the key functions of the transcription factor p53 is the regulation of the programmed cell death. In response to stress signals, the accumulation of p53 in the nucleus occurs in order to exert its proapoptotic role [61]. Its phosphorylation at numerous sites, comprising Ser15, Ser20, and Ser46, improves its transcriptional as well as proapoptotic ability. Thus, the triggering of p53 can up regulate or down regulate numerous target genes, whose promoters have consensus p53 response elements, such as Bcl-2 proteins, other transcription factors, and many molecules which, then, can initiate a sequence of actions leading to cell death [62]. Surely, an important link between p53-mediated activation and programmed cell death comes from its capability to modulate the transcription of proapoptotic proteins of the Bcl-2 group, including the multidomain protein Bax [63] and the BH3-only members Bid, phorbol-12-myristate-13-acetate-induced protein 1 (pmaip1/noxa), and Bcl-2 binding component 3 (bbc3/puma) [64 - 66]. In opposition, the promoter of the antiapoptotic protein Bcl-2 presents a p53-negative response element, suggesting that Bcl-2 may be a straight target of p53-mediated repression [63]. Furthermore, p53 transcriptionally triggers other different genes that have been associated with programmed cell death, for example PERP (p53 apoptosis effector related to PMP-22) [67]. The clear interpretation of the function of the genes implicated in p53-dependent apoptosis needs the investigations in a system suitable to the intricacy of the p53 apoptotic effect, which is performed through dissimilar collections of effectors agreeing to the cellular setting. In addition to these transcriptional tasks, some studies proposed that p53 may stimulate transcription-independent apoptosis [68]. Nevertheless, it has not been fully clarified how these members act downstream p53 to regulate programmed cell death and contribute to a net result that induces apoptosis in skeletal muscle tissue.

Numerous studies have engaged microarray knowledge to reveal variations in the expression of genes that go with aging in skeletal muscle of mice [69], rats [70], monkeys [71] and humans [72, 73]. Taking together, results acquired from these researches indicate that the expression of genes suggestive of cellular injury, such as those implicated in the stress or inflammatory response, rise with aging, while expression of metabolic and biosynthetic genes diminish with age. Microarray studies of young and ancient C57BL/6NHsd mice support earlier conclusions on specific transcript classes differentially expressed with age and points to increased p53 activity in muscle tissue of elder animals, may endorse to aging phenotypes [74]. The transcripts that are triggered by p53, or whose gene products are recognized to link to p53, significantly increase in elderly, whereas the expression of genes that are known to prevent p53 activity decay in ancient muscle. Two gene products that are known to have a critical function in the initiation of cell cycle arrest following DNA injury, cyclin-dependent kinase inhibitor 1A (p21) and growth arrest and DNA-damage-inducible, alpha (GADD45α), were both suggestively increased in the skeletal muscle of ancient mice [74]. As mentioned, a critical role of 'the guardian of the genome' is the initiation of apoptosis in cells that suffer DNA injury or reveal uncontrolled cell growth. The increased activation of p53 in aged skeletal muscle of C57BL/6NHsd mice is related to a significant increment in mRNA levels of genes known to be relevant transducers of p53-mediated cell death: Puma, Noxa, tumor necrosis factor receptor superfamily, member 10b (tnfrsf10b/killer/dr5) [75] and Bcl-2-related ovarian killer polypeptide (Bok) [76]. Thus, genes associated with the activation of apoptosis or the response to genomic damage, have increased mRNA levels in the older animals. The age-associated increment in genes that are stimulated during cellular stress and injury is possibly in response to an extensive diversity of biological impairments identified to accumulate in older muscle over the course of the lifetime [77].

Probably, increased oxidative stress may play a role in p53-mediated apoptosis in ancient muscle. Oxidative stress triggered by the action of H_2O_2 induces p53 activation in C2C12 murine skeletal muscle cells in a time-dependent way with a maximum level of activation and nuclear localization at 1–2 h of treatment, since active p53 moves from the cytosol to the nucleus to control its target genes [78]. However, the preincubation with physiological concentrations of testosterone or estradiol, inhibit the activation/ phosphorylation of p53 induced by the apoptotic agent, as it has been also observed in cells of cardiac muscle [79]. Probably, the stressful oxidative state to which myoblasts cultures are subjected with H_2O_2 treatment, is mitigated in the presence of the hormones and the environmental conditions probably become fewer adverse, being the stress stimulus, that induces p53 activation, less intense. These data sustenance the idea that sex steroids protect skeletal myoblasts against apoptosis by impeding the activation of the transcription factor p53 at the nucleus [74].

Modulation of Pro and Anti-Apoptotic p53-Related Genes by Sex Steroids during Oxidative Stress-Induced Apoptosis in Skeletal Muscle

The identification of numerous proapoptotic p53 targets, some of them that constrain antiapoptotic Bcl-2 family memberships, suggests that it is only through the combined transcriptional activation of several proapoptotic targets that p53 applies its wholly apoptotic potential. The initial mechanistic linking between p53 and Bcl-2 family came from the detection that p53 directly triggers the transcription of the apoptotic agent Bax [80]. Nevertheless, since this p53-mediated transactivation does not completely explain the capacity of p53 to block Bcl-2 antiapoptotic members and thus induce programmed cellular death, this can be explained by the action of proapoptotic BH3-only Bcl-2 family members. Several of these proteins, including Puma and Noxa, were recognized as transcriptional targets of p53. While Noxa has usually a slight activity, Puma is considered a crucial member of the apoptotic effects of p53, whose expression can be activated quickly in response to many types of stress in diverse tissues [81]. According to the activation of p53 triggered by oxidative stress in skeletal muscle, several genes have been reported to participate in the apoptotic process of the tissue, being reported sex steroids key regulators of this process in skeletal muscle.

PUMA and NOXA

Instigation of programmed cellular death necessitates not only pro-apoptotic Bcl-2 family members but also other molecules that are linked only by the small BH3 protein-interaction region. The BH3-only proteins are custodians that sense developmental death cues or intracellular injury. These proteins are expressed and activated by cellular stress circumstances and are conjectured to stimulate the mitochondrial outer membrane permeabilization by interrelating with the non-apoptotic Bcl-2 proteins or activating the oligomerization and pore-forming function of the pro-apoptotic Bcl-2 effector proteins, Bak and Bax [82 - 84]. Puma and Noxa are among the apoptotic p53-target genes whose products can confine at the mitochondria and function as a facilitator of p53-dependent apoptosis through the intrinsic pathway. Both of them share homology with Bcl-2 family members, but only within a small stretch of amino acids named the BH3 region (Bcl-2 homology 3). Different from Noxa, that usually has less strong activity, Puma is very effective in promoting cell death. When expressed, it destroys tumor cells within a few hours, and gene knockouts in human colorectal tumoral cells demonstrated that Puma was essential for cell death triggered by p53, hypoxia and DNA-damaging agents [84, 85, 86] being Puma a critical intermediary of programmed cell death in response to p53. Even though Noxa is a primary p53-response gene, in several tissues, induction of Noxa occurred well also in absence/deficiency of p53 showing that, Noxa activation in response to DNA damage can also arise in a p53-independent manner [87].

In C2C12 murine cells, the apoptosis induction with H_2O_2 leads to the increase of Puma mRNA levels with a maximum expression at 1 h. From then on, Puma mRNA decreased but preserved its levels over the control. However, no significant difference respect to control is observed in the transcript levels of Noxa gene, implying that Puma, but not Noxa, is transcriptionally controlled in H_2O_2-induced skeletal muscle apoptosis [88]. Given that p53 and Puma show the same pattern of activation/increased expression after apoptosis induction, it could be supposed that Puma is under the transcriptional regulation of p53, contributing to the myoblasts apoptosis. Pretreatment with testosterone or 17β-estradiol negatively regulates Puma mRNA levels, offsetting the apoptotic function of this protein of the Bcl-2 family and contributing thus to the protective action of sex hormones on skeletal muscle [77, 88]. The binding of Puma to the inhibitory proteins of the Bcl-2 group *via* its BH3 region looks to be a key regulatory phase in the promotion of programmed cell death. It results in the shift of the proteins Bax and/or Bak [85], followed by their activation and the creation of pores on the membrane of the mitochondria, which results in permeabilization of the organelle with the release of proapoptotic factors, conducing to mitochondrial dysfunction and caspase activation. It has been reported that both testosterone and 17β-estradiol reduce the levels of Bax expression and avoids the loss of mitochondrial membrane potential [89], so these findings further ascribe another mitochondrial defensive effect to sex steroids, by decreasing the transcription levels of Puma mRNA.

PERP

PERP (p53 apoptosis effector related to PMP-22), signifies a new kind of effector involved in p53-dependent apoptosis. The transcriptional stimulation of PERP by p53 seems to be vital for PERP′s capacity to trigger apoptosis and its overexpression is enough to induce programmed death in various cells types. Moreover, the PERP gene is greatly expressed in cells suffering p53-dependent apoptosis as compared to cells undergoing p53-dependent G1 arrest. While its particular mechanism of action has not been clarified, it is known that it functions merely to induce death and not cell cycle arrest [90, 91], being its requisite dictated by the cellular nature and setting [90]. Even though it has been exclusively linked to apoptosis, its specific function in provoking an apoptotic reaction has not been fully clarified.

In C2C12 skeletal muscle cells, the gene encoding PERP is up regulated by H_2O_2 -induced apoptosis with a extreme expression at 3 h of treatment, from then on the levels of PERP mRNA diminished up to control. In addition, treatment with physiological concentrations of 17β-estradiol or testosterone before the induction of apoptosis conduces to reduced levels of PERP mRNA, showing that sex hormones may exert its protective influence against apoptosis in skeletal muscle, down regulating this apoptotic factor at the transcriptional level, counteracting the effect of the oxidative stress [77, 88]. Furthermore, it could be proposed that p53 would be responsible for PERP transcription since they share the array of activation/increased expression. Thus, sex hormones are probably regulating this effect by obstruction of the p53 transcription activity, formerly described [77, 92].

BCL-2

Gene expression that controls the arrangement of programmed cell death, such as Bcl-2 which induces survival or the apoptotic genes Bax, Bim, or Noxa, show a crucial role in the determination of this survival threshold and will decide for cell survival or death next to injury [93]. Bcl-2 protein was the first intracellular regulator of programmed cell death to be recognized [94], and it has been broadly reported that high levels improve cellular subsistence under different cytotoxic situations [95, 96], inducing its overexpression the inhibition of apoptotic cell death [93]. Additionally to p53 ability to stimulate the transcription of Bcl-2 antagonists, this transcription factor uses additional approaches to control Bcl-2. It has been exposed that p53 can inhibit its transcription in some cells, while the pathway for such repression has not been wholly revealed. Nevertheless, it has been revealed the presence of a negative response element in the Bcl-2 gene through which p53 may either directly or indirectly, transcriptionally down regulate the expression of this gene implicated in apoptosis [63].

In C2C12 cell line, oxidative stress induced by hydrogen peroxide promotes the time-dependent down regulation of Bcl-2 gene expression, getting their lower levels afterward 4 h of incubation with the apoptotic inductor. When cells are pretreated with physiological concentrations of testosterone or 17β-estradiol before the apoptotic stimulus, the Bcl-2 mRNA levels increase above the corresponding apoptotic condition, counteracting the hormones the effects of the apoptotic agent. Thus, the antiapoptotic effect of sex steroids in C2C12 muscle cells includes the positive regulation of Bcl-2 at its transcription level [77, 88]. Of relevance, the transcription of the gene encoding this polypeptide is regulated by promoters P1 and P2 [97], which do not contain estrogen response elements (EREs) [98]. As EREs, usually located in the promoters, have also been observed in other domains [99], it could be proposed that E2 exerts its survival role through its interaction with these unusual EREs. Certainly, it has been showed that estradiol prompts the transcription of Bcl-2 through EREs not present in the promoters in MCF-7 cells [98].

BIM

Bim is another p53 target gene, belonging to the group of the BH3-only subset of the Bcl-2 family proteins, which plays an important function triggering the cell death. It can interrelate with some, but not all, Bcl-2 family members that promote cell survival, and only those antiapoptotic molecules that link to it, can counteract its proapoptotic action [100]. Undoubtedly, it has been demonstrated that Bim is able to link with Bcl-2, avoiding apoptosis [101]. Although numerous theories have been suggested to describe the cellular concerns of the interactions between antiapoptotic and BH3-only members of Bcl-2 group, none of them explains one perplexing aspect: the net result of their connections, such as occurs among Bim and Bcl-2 [102]. It has been proposed that the ultimate outcome, would be severely linked to the cell type and the apoptosis inducer.

Bim is transcriptionally up regulated in H_2O_2 -induced apoptosis in murine skeletal muscle cells, but opposing to what it is expected, testosterone or 17β-estradiol pretreatment before H_2O_2 treatment, cannot reduce Bim transcription levels. Instead, no significant changes on mRNA Bim levels are detected respect to apoptotic induction treatment [77, 88]. Given that the ability of Bcl-2 to interact with Bim, the results obtained in these works, where it was showed that the apoptosis inducer augmented the mRNA transcript level of Bim at the same time that reduced the transcript level of Bcl-2, upkeep the idea that it would prevail the apoptotic role of Bim. Since under these conditions, Bim could not be totally appropriated by Bcl-2 to inhibit its apoptotic function, the net effect would be the promotion of apoptosis. In contrast, even though treatment with sex hormones previous apoptosis induction does not induce significant changes on mRNA Bim levels to respect to apoptotic induction treatment, at the same time they induce an increase in mRNA transcript levels of Bcl-2, signifying that owing to the interaction of the resultant proteins, it would be blocked the apoptotic role of Bim.

P66Shc

One key consequence of the genetic program elicited by p53 throughout the induction of cell death is the growth of the ROS produced by the mitochondria [103]. p53 positively controls the expression of p66Shc, a mitochondrial producer of H_2O_2. Activated p53 enhances p66Shc stability, and it is incapable to triggers apoptosis in p66Shc-/- fibroblasts, proposing that p66Shc controls p53-linked apoptosis [104]. The activation of the β-adrenergic receptors in differentiated H9c2 cardiomyoblasts, leads to cell death and cardiovascular worsening, with an increment in the levels of p53 and phosphorylated-p66Shc [105]. Nevertheless, the molecular process through which p66Shc arbitrates oxidative stress-induced programmed cell death continue basically unidentified.

The mammalian adaptor molecule ShcA is presented in different isoforms, p46Shc, p52Shc, and p66Shc, derived from a single gene through the varied use of transcription/translation initiation regions and alternative splicing. The p66Shc isoform has an extra amino-terminal collagen homology-like region (CH2), unlike p46 and p52, which encloses a serine residue at site 36 (Ser36), that is phosphorylated in response to numerous stimuli such as H_2O_2 and UV irradiation [106]. While p52 and p46 are cytoplasmic signal transducers implicated in the mitogenic cascade from activated tyrosine kinase receptors to Ras, the p66 isoform orders the metabolism of the reactive oxygen species and cell death [104, 106 - 108]. Similarly, p66Shc protein exerts an important inhibitory signaling effect on the ERK signal cascade in skeletal myoblasts, which is necessary for the proper modulation of actin cytoskeleton organization and the typical glucose transport responses, contributing to the variations in glucose uptake [109, 110].

P66Shc protein is related to ROS signaling whose expression differs between dissimilar cellular types and even is not expressed in some cells [107,108]. It has come into focus as a main factor of cellular vulnerability to ROS damage controlling mammalian life span by modulating the cellular reaction to oxidative injury [106]. Its expression and phosphorylation at serine 36 are significant for the apoptotic response upon oxidative damage during elderly [106, 111].

It has been showed that, at the transcriptional level, both testosterone and 17β-estradiol are able to reverse the up regulation of mRNA p66Shc triggered by the apoptotic agent H_2O_2 in skeletal muscle [53, 92]. Phosphorylation in serine 36 is a critical phase in the p66Shc reaction to ROS injury and its apoptotic role. Reliant on cellular context and the type of the signal, p66Shc can be phosphorylated by diverse kinases, such as PKC and JNK [112,113], and migrates to mitochondria where it causes alterations in the structure and physiology of the organelle, producing deleterious molecules, provoking the mPTP opening and the release of apoptotic factors such as cytochrome c [112, 114]. Accordingly, sex steroids can also reduce the activation/phosphorylation of the p66Shc protein, triggered by hydrogen peroxide in C2C12 skeletal muscle cells. The hormonal treatment previous to the death signal of H_2O_2 is able to diminish the levels of p-p66Shc and its mitochondrial localization, augmented next the induction of apoptosis [53,92]. Oxidative insult induces, in myoblasts, the activation of p53, probably contributing thus to the rise of circulating levels of p66Shc.

MDM2

p53 levels are the single most relevant factor of its role, being through the turnover of the protein, the key manner by which p53 levels are regulated. In healthy unstressed cells, p53 is a highly unstable polypeptide with a short half-life fluctuating from 5 to 30 min, which is present at very low intracellular levels owed to incessant demolition mostly mediated by MDM2 [115], whereas under stress circumstances its half-life is significantly sustained. MDM2 is the principal opponent of p53 action, that both blocks p53 transcriptional activity directly (sterically) and arbitrates its destruction through an ubiquitin-dependent *via* on nuclear and cytoplasmic 26S proteasomes [115]. MDM2 itself is also a transcriptional target of p53, which generates an auto regulatory negative loop whereby p53 coordinates the expression of its own negative regulator, keeping low intracellular p53 levels in the nonexistence of stress. Additionally, there are extra controls that regulate p53 function through post-translational modifications, such as phosphorylation in serine 15, which results in the disturbance of the p53-MDM2 complex, leading to the increase of active p53 in the cell [116].

In myoblasts, the regulation of mRNA transcript levels of MDM2 rest on the moment of the apoptotic process assessed. At the beginning of apoptosis induction (30 min^{-1} h treatment with the apoptotic agent H_2O_2), MDM2 transcription level is up regulated, in order to promote its new protein synthesis, thus enhance its interaction with p53 and the subsequent destruction of the transcription factor. The increment of MDM2 mRNA levels in the first phase of apoptosis in C2C12 cells could be part of the protection reaction, formerly described [55] to protect cells from apoptosis, by stimulating the destruction of p53. Nevertheless, at longer treatment periods (3-4 h, H_2O_2) the diminution in MDM2 mRNA transcript to control condition, arises. In advanced phases of programmed cell death, the predominant action of p53 transcription factor has taken place and gotten the cells into the apoptotic program. The MDM2 levels go down to basal condition and, as a consequence, to the reduction in the rate of p53 elimination. Pretreatment with physiological concentrations of sex hormones up regulates MDM2 transcript level, supported by this manner the ubiquitination and degradation of p53, showing its antiapoptotic function in skeletal muscle [77, 88].

Modulation of Trpm-2/Clusterin during Castration-Induced Apoptosis in Levator Ani Skeletal Muscle

The testosterone-repressed prostate message-2 (trpm-2) or clusterin, was firstly isolated and cloned from the regressing ventral prostate of the rat [117]. In this localization, and in other tissues responsive to hormone such as the mammary gland, this gene is stimulated when there is a deficit of the appropriate trophic hormone. Clusterin is an unknowable glycoprotein that is nearly ubiquitous tissue distributed, seemingly implicated in various biological processes such as sperm maturation, tissue differentiation and remodeling, membrane reprocessing, lipid transport, cell-cell or cell-substratum communication, cellular proliferation, and cell death. Trpm-2 is also supposed to be implicated in several pathological conditions such as neurodegeneration, elderly and cancer [118 - 121]. Of relevance, its expression provides a significant and early indicator of apoptosis in many types of mammalian cells [122]. The trpm-2 gene is expressed in cells resisting apoptosis and is therefore considered as an anti-apoptotic gene [123 - 125]. Trpm-2/clusterin expression has been detected in skeletal muscle [126] but its role in muscle response to the apoptotic signal is yet to be investigated.

The overexpression of the secretory Clusterin isoform defends the cell from programmed cell death triggered by cellular stress, such as chemo- and radiotherapy, or sex steroids diminution. Clusterin stimulates cellular survival by a quantity of means, that comprise blockage of BAX at the mitochondrial level, regulation of ERK1/2 cascade and matrix metallopeptidase-9 expression, activation of the phosphatidylinositol 3-kinase/protein kinase B cascade, stimulation of angiogenesis, and regulation of the nuclear factor kappa B (NF-κB) *via*. On the other hand, its down regulation allows p53 activation, which conduces to the skew of the proapoptotic/antiapoptotic ratio of Bcl-2 family members, leading to mitochondrial dysfunction and apoptosis [127]. p53 may also transcriptionally block secretory Clusterin to further induce the proapoptotic pathway since the loss of functional p53 in HCT116:p53(−/−) cells increased Clusterin *de novo* synthesis after IR treatment. Suppression of secretory Clusterin protein levels by p53 may be central for the pathway of p53-mediated events resulting in apoptosis afterward IR exposure or treatment with other cytotoxic agent [128].

It has been demonstrated that apoptosis takes part in the castration-induced atrophy of the levator ani (LA) muscle [129]. The LA muscle of the male rat is exquisitely subtle to androgens much like the prostate or the seminal vesicles [130,131]. In contrast to a typical skeletal muscle [132], castration induces severe atrophy of the LA resulting in significant ultrastructural variations and alterations in its contractile properties [133]. Gonadectomy of male rats conduces to the overexpression of trpm-2/clusterin in the LA muscle that returned to control level upon androgen replacement, suggesting that an anti-apoptotic response is being triggered in response to androgen withdrawal, probably as a defense response against apoptosis. In contrast, androgen withdrawal does not alter the trpm-2 expression in the superficial vastus lateralis muscle. These results indicate that the androgen response of the sexually dimorphic LA muscle closely resembles that of a typical hormone responsive tissue since obvious features of programmed cell death are detected as a consequence of androgen withdrawal [129].

Estradiol Regulation of miRNAs Expression Associated with Skeletal Muscle Apoptosis

Noncoding RNAs and post-transcriptional regulation of miRNA sets a supplementary layer of control over gene expression. miRNAs are small molecules of 21-23 nucleotides transcribed either independently or from the intronic domains of protein-coding genes. The primary transcribed miRNA is sequentially handled by Drosha to a hairpin, which is transported to the cytoplasm and further processed into a mature miRNA by Dicer. They are implicated in post-transcriptional gene regulation by linking to the 3′UTR site of their mRNAs target and in blocking translational initiation leading to diminished gene expression. Knockdown of the Dicer gene in mice exposed the significance of miRNAs in skeletal muscle growth. Both embryonic and postnatal muscle development are affected by the lack of miRNAs producing hypoplasia and death [134]. Small RNAs are vital constituents in the net of skeletal muscle development with very precise roles in proliferation, differentiation, and maturation of satellite cells. miR-1, miR-133, and miR-206 are fine investigated muscle-specific miRNAs [135 - 140]. They coordinately work with the myogenic regulatory factors (MRFs), which have a main role in the modulation of myogenic development, and control cell fate. miR-1 promotes skeletal muscle differentiation, while miR-133 regulates proliferation. Numerous other miRNAs have also been identified for their fundamental functions in myogenesis.

The importance of miRNAs in myosatellite cell maintenance [141, 142] proliferation [143, 144] and differentiation [145, 146] is well studied in diverse physiological and disease situations of muscle [147 - 150]. The functions of miRNAs during different metabolic phases of muscle counting cell death [151, 152] and hypertrophy [153, 154, 158] have also been determined. The contribution of miRNAs in modulating different signaling pathways [153 - 156] implicated in myogenesis states their diverse functions in any living cell.

17β-Estradiol negatively affects muscle growth in rainbow trout. In salmonids, the effect of 17β-Estradiol during ovarian growth contributes to the loss in muscle mass or reductions in muscle growth [157 - 159] contrasting rodent, bovine and porcine species in which 17β-Estradiol has anabolic properties. It has been observed that 17β-Estradiol can influence procedures affecting myogenesis and muscle growth, partly through regulation of miRNA-related mechanisms in rainbow trout skeletal muscle. There has been recognized 36 differentially expressed miRNAs including those which participates in myogenesis, cell cycle, and cell death. Some central myogenic miRNAs, such as miR-133 and miR-206, are positively regulated but others like miR-145 and miR-499, are down regulated by estradiol treatment [160]. Of relevance, the expression of omy-miR-23a-3p is down regulated in rainbow trout skeletal muscles under the influence of 17β-Estradiol. Gene ontology enrichment of Omy-miR-23a-3p exposed its function in modulating mitochondrial outer membrane permeability linking thus its function in programmed cell death. Thirteen target genes of this miRNA control mitochondrial organization, regulation, and outer membrane permeabilization, suggesting the role of mitochondrial permeability in muscle cells under the effect of female hormone. As the downstream effects of mitochondrial outer membrane permeability, the expression of caspases genes, the expression of the initiator caspase 9 and executor caspase 3 are higher in estradiol-treated samples matched to the control, whereas the expression of initiator caspase 8 does not display any difference [160]. Therefore, additional underwriting to muscle catabolism may be increased in mitochondrial outer membrane permeability (MOMP), the integrity of which is sustained by Omy-miR-23a-3p [161]. Thus, E2-induced diminutions in miR-23a-3p may increase MOMP, a mechanism linked with caspase-induced cellular death [161], which themselves are regulated by estradiol treatment in rainbow trout. These results suggest that sex hormones can alter processes affecting myogenesis and muscle growth, partly through regulation of miRNA-

ting myogenesis and muscle growth, partly through regulation of miRNA-related mechanisms in skeletal muscle.

CONCLUDING REMARKS

Evidence shows that skeletal muscle apoptosis occurs under a variety of conditions. However, in general, the literature associate the apoptosis of skeletal muscle with pathological conditions as sarcopenia, atrophy, ischemia followed by reperfusion, accelerated aging and injury due to exercise.

The data presented in this chapter summarizes the generalities of skeletal muscle at the molecular/structural and functional level and basic concepts of apoptosis. Which it focuses specifically, at a molecular level, on the actions of both sex steroids in the homeostasis of skeletal muscle tissue in physiological and pathological conditions, focusing on the relationship of muscle apoptosis and sexual hormones. Sex hormones exert an antiapoptotic effect by blocking the effect of the apoptotic stimulus in the regulation of the transcription of genes implicated in different phases of the apoptosis. The transcription factor p53 plays a key role in the modulation of apoptosis in skeletal muscle, by the activation of different conjugated and simultaneous mechanisms that promote the regulation of proapoptotic genes as reaction to stress. Altogether, these data will support the significant role of 17β-Estradiol and testosterone in the abrogation of several cellular signaling paths in skeletal myoblast cells that act in concert conduce to cellular death. Therefore, this summery could afford details of the molecular mechanisms elicited by sex hormones during muscle pathologies that result in the loss of muscle mass, related to endocrine misbalance.

CONSENT FOR PUBLICATION

Not applicable.

CONFLICT OF INTEREST

The authors confirm that this chapter contents have no conflict of interest.

ACKNOWLEDGEMENTS

National University of the South Argentina and National Research Council of Argentina (CONICET).

REFERENCES

[1]   Van De Graaff K. Muscular System. Human Anatomy 6th McGraw-Hill Higher Education 2002; pp. 233-50.

[2]   Sweeney HL, Hammers DW. Muscle Contraction. Cold Spring Harb Perspect Biol 2018; 10(2): 1-13.

[http://dx.doi.org/10.1101 / cshperspect.a023200] [PMID: 294194 05]

[3]   Baldwin KM, Haddad F. Effects of different activity and inactivity paradigms on myosin heavy chain gene expression in striated muscle. J Appl Physiol 2001; 90(1): 345-57.
[http://dx.doi.org/10.1152/jappl.2001.90.1.345] [PMID: 11133928]

[4]   van Rooij E, Quiat D, Johnson BA, *et al.* A family of microRNAs encoded by myosin genes governs myosin expression and muscle performance. Dev Cell 2009; 17(5): 662-73.
[http://dx.doi.org/10.1016/j.devcel.2009.10.013] [PMID: 19922871]

[5]   Sharma S, Eghbali M. Influence of sex differences on microRNA gene regulation in disease. Biol Sex Differ 2014; 5(3) 6410-5-3
[http://dx.doi.org/10.1186/2042-6410-5-3]

[6]   Knöll R, Hoshijima M, Chien K. Z-line proteins: implications for additional functions. Eur Heart J Suppl 2002; 4: I13-7.
[http://dx.doi.org/10.1016/S1520-765X(02)90105-7]

[7]   Huxley AF, Niedergerke R. Structural changes in muscle during contraction; interference microscopy of living muscle fibres. Nature 1954; 173(4412): 971-3.
[http://dx.doi.org/10.1038/173971a0] [PMID: 13165697]

[8]   Hopkins PM. Voluntary motor systems—skeletal muscle, reflexes, and control of movement. Foundations of Anesthesia. 2nd ed., London: Mosby 2005.

[9]   Evans WJ. What is sarcopenia? J Gerontol 1995; 50: 5-8.

[10]   Rantanen T, Avlund K, Suominen H, Schroll M, Frändin K, Pertti E. Muscle strength as a predictor of onset of ADL dependence in people aged 75 years. Aging Clin Exp Res 2002; 14(3) (Suppl.): 10-5.
[PMID: 12475129]

[11]   Sinha-Hikim I, Cornford M, Gaytan H, Lee ML, Bhasin S. Effects of testosterone supplementation on skeletal muscle fiber hypertrophy and satellite cells in community-dwelling older men. J Clin Endocrinol Metab 2006; 91(8): 3024-33.
[http://dx.doi.org/10.1210/jc.2006-0357] [PMID: 16705073]

[12]   Kelley DE, Goodpaster B, Wing RR, Simoneau JA. Skeletal muscle fatty acid metabolism in association with insulin resistance, obesity, and weight loss. Am J Physiol 1999; 277(6): E1130-41.
[PMID: 10600804]

[13]   Iizuka K, Machida T, Hirafuji M. Skeletal muscle is an endocrine organ. J Pharmacol Sci 2014; 125(2): 125-31.
[http://dx.doi.org/10.1254/jphs.14R02CP] [PMID: 24859778]

[14]   Schnyder S, Handschin C. Skeletal muscle as an endocrine organ: PGC-1α, myokines and exercise. Bone 2015; 80: 115-25.
[http://dx.doi.org/10.1016/j.bone.2015.02.008] [PMID: 26453501]

[15]   Aizawa K, Iemitsu M, Maeda S, *et al.* Expression of steroidogenic enzymes and synthesis of sex steroid hormones from DHEA in skeletal muscle of rats. Am J Physiol Endocrinol Metab 2007; 292(2): E577-84.
[http://dx.doi.org/10.1152/ajpendo.00367.2006] [PMID: 17018772]

[16]   Sato K, Iemitsu M, Aizawa K, Ajisaka R. Testosterone and DHEA activate the glucose metabolism-related signaling pathway in skeletal muscle. Am J Physiol Endocrinol Metab 2008; 294(5): E961-8.
[http://dx.doi.org/10.1152/ajpendo.00678.2007] [PMID: 18349113]

[17]   Sato K, Iemitsu M. Exercise and sex steroid hormones in skeletal muscle. J Steroid Biochem Mol Biol 2015; 145: 200-5.
[http://dx.doi.org/10.1016/j.jsbmb.2014.03.009] [PMID: 24704257]

[18]   Glenmark B, Nilsson M, Gao H, Gustafsson JA, Dahlman-Wright K, Westerblad H. Difference in skeletal muscle function in males *vs.* females: role of estrogen receptor-beta. Am J Physiol Endocrinol

Metab 2004; 287(6): E1125-31.
[http://dx.doi.org/10.1152/ajpendo.00098.2004] [PMID: 15280152]

[19]   Lee WJ, Thompson RW, McClung JM, Carson JA. Regulation of androgen receptor expression at the onset of functional overload in rat plantaris muscle. Am J Physiol Regul Integr Comp Physiol 2003; 285(5): R1076-85.
[http://dx.doi.org/10.1152/ajpregu.00202.2003] [PMID: 14557238]

[20]   Brown M. Skeletal muscle and bone: effect of sex steroids and aging. Adv Physiol Ed 2008; 32:120–6.

[21]   Tiidus PM. Benefits of estrogen replacement for skeletal muscle mass and function in post-menopausal females: evidence from human and animal studies. Eur J Med 2011; 43(2): 109-14.
[http://dx.doi.org/10.5152/eajm.2011.24] [PMID: 25610174]

[22]   Vasconsuelo A, Milanesi L, Boland R. Actions of 17β-estradiol and testosterone in the mitochondria and their implications in aging. Ageing Res Rev 2013; 12(4): 907-17.
[http://dx.doi.org/10.1016/j.arr.2013.09.001] [PMID: 24041489]

[23]   La Colla A, Pronsato L, Milanesi L, Vasconsuelo A. 17β-Estradiol and testosterone in sarcopenia: Role of satellite cells. Ageing Res Rev 2015; 24(Pt B): 166-77.
[http://dx.doi.org/10.1016/j.arr.2015.07.011] [PMID: 26247846]

[24]   Reid KF, Fielding RA. Skeletal muscle power: a critical determinant of physical functioning in older adults. Exerc Sport Sci Rev 2012; 40(1): 4-12.
[http://dx.doi.org/10.1097/JES.0b013e31823b5f13] [PMID: 22016147]

[25]   Kerr JF, Wyllie AH, Currie AR. Apoptosis: a basic biological phenomenon with wide-ranging implications in tissue kinetics. Br J Cancer 1972; 26(4): 239-57.
[http://dx.doi.org/10.1038/bjc.1972.33] [PMID: 4561027]

[26]   Kanduc D, Mittelman A, Serpico R, *et al.* Cell death: apoptosis *versus* necrosis (review). Int J Oncol 2002; 21(1): 165-70.
[PMID: 12063564]

[27]   Vanlangenakker N, Vanden Berghe T, Vandenabeele P. Many stimuli pull the necrotic trigger, an overview. Cell Death Differ 2012; 19(1): 75-86.
[http://dx.doi.org/10.1038/cdd.2011.164] [PMID: 22075985]

[28]   Kar B, Sivamani S. Apoptosis: Basic Concepts, Mechanisms and Clinical Implications. IJPSR 2015; 6: 940-50.

[29]   Thornberry NA, Lazebnik Y. Caspases: enemies within. Science 1998; 281(5381): 1312-6.
[http://dx.doi.org/10.1126/science.281.5381.1312] [PMID: 9721091]

[30]   Julien O, Wells JA. Caspases and their substrates. Cell Death Differ 2017; 24(8): 1380-9.
[http://dx.doi.org/10.1038/cdd.2017.44] [PMID: 28498362]

[31]   McArthur K, Kile BT. Apoptotic Caspases: multiple or mistaken identities? Trends Cell Biol 2018; 28(6): 475-93. Epub ahead of print
[http://dx.doi.org/10.1016/j.tcb.2018.02.003] [PMID: 29551258]

[32]   Brown JM, Attardi LD. The role of apoptosis in cancer development and treatment response. Nat Rev Cancer 2005; 5(3): 231-7.
[http://dx.doi.org/10.1038/nrc1560] [PMID: 15738985]

[33]   Shi Y. Mechanisms of caspase activation and inhibition during apoptosis. Mol Cell 2002; 9(3): 459-70.
[http://dx.doi.org/10.1016/S1097-2765(02)00482-3] [PMID: 11931755]

[34]   Cosentino K, García-Sáez AJ. Mitochondrial alterations in apoptosis. Chem Phys Lipids 2014; 181: 62-75.
[http://dx.doi.org/10.1016/j.chemphyslip.2014.04.001] [PMID: 24732580]

[35]   Liu Q, Wang HG. Anti-cancer drug discovery and development: Bcl-2 family small molecule

inhibitors. Commun Integr Biol 2012; 5(6): 557-65.
[http://dx.doi.org/10.4161/cib.21554] [PMID: 23336025]

[36] Zou H, Henzel WJ, Liu X, Lutschg A, Wang X. Apaf-1, a human protein homologous to *C. elegans*
CED-4, participates in cytochrome c-dependent activation of caspase-3. Cell 1997; 90(3): 405-13.
[http://dx.doi.org/10.1016/S0092-8674(00)80501-2] [PMID: 9267021]

[37] Li P, Nijhawan D, Budihardjo I, *et al.* Cytochrome c and dATP-dependent formation of Apaf-
1/caspase-9 complex initiates an apoptotic protease cascade. Cell 1997; 91(4): 479-89.
[http://dx.doi.org/10.1016/S0092-8674(00)80434-1] [PMID: 9390557]

[38] Boatright KM, Renatus M, Scott FL, *et al.* A unified model for apical caspase activation. Mol Cell
2003; 11(2): 529-41.
[http://dx.doi.org/10.1016/S1097-2765(03)00051-0] [PMID: 12620239]

[39] Always E. Apoptosis in Skeletal Muscle Health and Conditions of Muscle Wasting. Apoptosis:
Modern Insights into Disease from Molecules to Man. 2010.
[http://dx.doi.org/10.1201/9781439845431-14]

[40] Always S, Siu P. Nuclear Apoptosis and Sarcopenia.Sarcopenia – Age-Related Muscle Wasting and
Weakness. 2011.
[http://dx.doi.org/10.1007/978-90-481-9713-2_9]

[41] Adhihetty PJ, Hood DA. Mechanisms of apoptosis in skeletal muscle. Basic Appl Myol 2003; 13:
171-9.

[42] Pronsato L, Boland R, Milanesi L. Non-classical localization of androgen receptor in the C2C12
skeletal muscle cell line. Arch Biochem Biophys 2013; 530(1): 13-22.
[http://dx.doi.org/10.1016/j.abb.2012.12.011] [PMID: 23262317]

[43] Milanesi L, Russo de Boland A, Boland R. Expression and localization of estrogen receptor alpha in
the C2C12 murine skeletal muscle cell line. J Cell Biochem 2008; 104(4): 1254-73.
[http://dx.doi.org/10.1002/jcb.21706] [PMID: 18348185]

[44] Milanesi L, Vasconsuelo A, de Boland AR, Boland R. Expression and subcellular distribution of
native estrogen receptor beta in murine C2C12 cells and skeletal muscle tissue. Steroids 2009; 74(6):
489-97.
[http://dx.doi.org/10.1016/j.steroids.2009.01.005] [PMID: 19428437]

[45] Spangenburg EE, Geiger PC, Leinwand LA, Lowe DA. Regulation of physiological and metabolic
function of muscle by female sex steroids. Med Sci Sports Exerc 2012; 44(9): 1653-62.
[http://dx.doi.org/10.1249/MSS.0b013e31825871fa] [PMID: 22525764]

[46] Bhasin S, Calof OM, Storer TW, *et al.* Drug insight: Testosterone and selective androgen receptor
modulators as anabolic therapies for chronic illness and aging. Nat Clin Pract Endocrinol Metab 2006;
2(3): 146-59.
[http://dx.doi.org/10.1038/ncpendmet0120] [PMID: 16932274]

[47] Sinha-Hikim I, Artaza J, Woodhouse L, *et al.* Testosterone-induced increase in muscle size in healthy
young men is associated with muscle fiber hypertrophy. Am J Physiol Endocrinol Metab 2002; 283(1):
E154-64.
[http://dx.doi.org/10.1152/ajpendo.00502.2001] [PMID: 12067856]

[48] Sinha-Hikim I, Roth SM, Lee MI, Bhasin S. Testosterone-induced muscle hypertrophy is associated
with an increase in satellite cell number in healthy, young men. Am J Physiol Endocrinol Metab 2003;
285(1): E197-205.
[http://dx.doi.org/10.1152/ajpendo.00370.2002] [PMID: 12670837]

[49] Brown TJ, Khan T, Jones KJ. Androgen induced acceleration of functional recovery after rat sciatic
nerve injury. Restor Neurol Neurosci 1999; 15(4): 289-95.
[PMID: 12671219]

[50] Boissonneault G. Evidence of apoptosis in the castration-induced atrophy of the rat levator ani muscle.

Endocr Res 2001; 27(3): 317-28.
[http://dx.doi.org/10.1081/ERC-100106009] [PMID: 11678579]

[51]   Vasconsuelo A, Milanesi L, Boland R. 17Beta-estradiol abrogates apoptosis in murine skeletal muscle cells through estrogen receptors: role of the phosphatidylinositol 3-kinase/Akt pathway. J Endocrinol 2008; 196(2): 385-97.
[http://dx.doi.org/10.1677/JOE-07-0250] [PMID: 18252962]

[52]   La Colla A, Vasconsuelo A, Boland R. Estradiol exerts antiapoptotic effects in skeletal myoblasts *via* mitochondrial PTP and MnSOD. J Endocrinol 2013; 216(3): 331-41.
[http://dx.doi.org/10.1530/JOE-12-0486] [PMID: 23213199]

[53]   La Colla A, Boland R, Vasconsuelo A. 17b-estradiol abrogates apoptosis inhibiting PKCd, JNK, and p66Shc activation in C2C12 cells. J Cell Biochem 2015; 116(7): 1454-65.
[http://dx.doi.org/10.1002/jcb.25107] [PMID: 25649128]

[54]   Pronsato L, Ronda AC, Milanesi LM, Vasconsuelo AA, Boland RL. Protective role of 17b-estradiol and testosterone in apoptosis of skeletal muscle. Actual Osteol 2010; 6: 66-80.

[55]   Pronsato L, Boland R, Milanesi L. Testosterone exerts antiapoptotic effects against H_2O_2 in C2C12 skeletal muscle cells through the apoptotic intrinsic pathway. J Endocrinol 2012; 212(3): 371-81.
[http://dx.doi.org/10.1530/JOE-11-0234] [PMID: 22219300]

[56]   Murray-Zmijewski F, Slee EA, Lu X. A complex barcode underlies the heterogeneous response of p53 to stress. Nat Rev Mol Cell Biol 2008; 9(9): 702-12.
[http://dx.doi.org/10.1038/nrm2451] [PMID: 18719709]

[57]   Levine AJ, Oren M. The first 30 years of p53: growing ever more complex. Nat Rev Cancer 2009; 9(10): 749-58.
[http://dx.doi.org/10.1038/nrc2723] [PMID: 19776744]

[58]   Hollstein M, Hainaut P. Massively regulated genes: the example of TP53. J Pathol 2010; 220(2): 164-73.
[PMID: 19918835]

[59]   Zhao R, Gish K, Murphy M, *et al.* Analysis of p53-regulated gene expression patterns using oligonucleotide arrays. Genes Dev 2000; 14(8): 981-93.
[PMID: 10783169]

[60]   Weinberg RL, Veprintsev DB, Bycroft M, Fersht AR. Comparative binding of p53 to its promoter and DNA recognition elements. J Mol Biol 2005; 348(3): 589-96.
[http://dx.doi.org/10.1016/j.jmb.2005.03.014] [PMID: 15826656]

[61]   Vousden KH, Lu X. Live or let die: the cell's response to p53. Nat Rev Cancer 2002; 2(8): 594-604.
[http://dx.doi.org/10.1038/nrc864] [PMID: 12154352]

[62]   Prives C, Hall PA. The p53 pathway. J Pathol 1999; 187(1): 112-26.
[http://dx.doi.org/10.1002/(SICI)1096-9896(199901)187:1<112::AID-PATH250>3.0.CO;2-3] [PMID: 10341712]

[63]   Miyashita T, Harigai M, Hanada M, Reed JC. Identification of a p53-dependent negative response element in the bcl-2 gene. Cancer Res 1994; 54(12): 3131-5.
[PMID: 8205530]

[64]   Oda E, Ohki R, Murasawa H, *et al.* Noxa, a BH3-only member of the Bcl-2 family and candidate mediator of p53-induced apoptosis. Science 2000; 288(5468): 1053-8.
[http://dx.doi.org/10.1126/science.288.5468.1053] [PMID: 10807576]

[65]   Nakano K, Vousden KH. PUMA, a novel proapoptotic gene, is induced by p53. Mol Cell 2001; 7(3): 683-94.
[http://dx.doi.org/10.1016/S1097-2765(01)00214-3] [PMID: 11463392]

[66]   Sax JK, Fei P, Murphy ME, Bernhard E, Korsmeyer SJ, El-Deiry WS. BID regulation by p53

contributes to chemosensitivity. Nat Cell Biol 2002; 4(11): 842-9.
[http://dx.doi.org/10.1038/ncb866] [PMID: 12402042]

[67]   Attardi LD, Reczek EE, Cosmas C, *et al.* PERP, an apoptosis-associated target of p53, is a novel member of the PMP-22/gas3 family. Genes Dev 2000; 14(6): 704-18.
[PMID: 10733530]

[68]   Caelles C, Helmberg A, Karin M. p53-dependent apoptosis in the absence of transcriptional activation of p53-target genes. Nature 1994; 370(6486): 220-3.
[http://dx.doi.org/10.1038/370220a0] [PMID: 8028670]

[69]   Lee CK, Klopp RG, Weindruch R, Prolla TA. Gene expression profile of aging and its retardation by caloric restriction. Science 1999; 285(5432): 1390-3.
[http://dx.doi.org/10.1126/science.285.5432.1390] [PMID: 10464095]

[70]   Sreekumar R, Unnikrishnan J, Fu A, *et al.* Effects of caloric restriction on mitochondrial function and gene transcripts in rat muscle. Am J Physiol Endocrinol Metab 2002; 283(1): E38-43.
[http://dx.doi.org/10.1152/ajpendo.00387.2001] [PMID: 12067840]

[71]   Kayo T, Allison DB, Weindruch R, Prolla TA. Influences of aging and caloric restriction on the transcriptional profile of skeletal muscle from rhesus monkeys. Proc Natl Acad Sci USA 2001; 98(9): 5093-8.
[http://dx.doi.org/10.1073/pnas.081061898] [PMID: 11309484]

[72]   Welle S, Brooks AI, Delehanty JM, *et al.* Skeletal muscle gene expression profiles in 20-29 year old and 65-71 year old women. Exp Gerontol 2004; 39(3): 369-77.
[http://dx.doi.org/10.1016/j.exger.2003.11.011] [PMID: 15036396]

[73]   Welle S, Brooks AI, Delehanty JM, Needler N, Thornton CA. Gene expression profile of aging in human muscle. Physiol Genomics 2003; 14(2): 149-59.
[http://dx.doi.org/10.1152/physiolgenomics.00049.2003] [PMID: 12783983]

[74]   Edwards MG, Anderson RM, Yuan M, Kendziorski CM, Weindruch R, Prolla TA. Gene expression profiling of aging reveals activation of a p53-mediated transcriptional program. BMC Genomics 2007; 8: 80.
[http://dx.doi.org/10.1186/1471-2164-8-80] [PMID: 17381838]

[75]   Wu GS, Burns TF, McDonald ER III, *et al.* KILLER/DR5 is a DNA damage-inducible p53-regulated death receptor gene. Nat Genet 1997; 17(2): 141-3.
[http://dx.doi.org/10.1038/ng1097-141] [PMID: 9326928]

[76]   Yakovlev AG, Di Giovanni S, Wang G, Liu W, Stoica B, Faden AI. BOK and NOXA are essential mediators of p53-dependent apoptosis. J Biol Chem 2004; 279(27): 28367-74.
[http://dx.doi.org/10.1074/jbc.M313526200] [PMID: 15102863]

[77]   La Colla A, Vasconsuelo A, Milanesi L, Pronsato L. 17b-Estradiol protects skeletal myoblasts from apoptosis through p53, Bcl-2, and Foxo families. J Cell Biochem 2017; 118(1): 104-15.
[http://dx.doi.org/10.1002/jcb.25616] [PMID: 27249370]

[78]   Lee CM, Lopez ME, Weindruch R, Aiken JM. Association of age-related mitochondrial abnormalities with skeletal muscle fiber atrophy. Free Radic Biol Med 1998; 25(8): 964-72.
[http://dx.doi.org/10.1016/S0891-5849(98)00185-3] [PMID: 9840742]

[79]   Liu H, Pedram A, Kim JK. Oestrogen prevents cardiomyocyte apoptosis by suppressing p38α-mediated activation of p53 and by down-regulating p53 inhibition on p38β. Cardiovasc Res 2011; 89(1): 119-28.
[http://dx.doi.org/10.1093/cvr/cvq265] [PMID: 20724307]

[80]   Chipuk JE, Kuwana T, Bouchier-Hayes L, *et al.* Direct activation of Bax by p53 mediates mitochondrial membrane permeabilization and apoptosis. Science 2004; 303(5660): 1010-4.
[http://dx.doi.org/10.1126/science.1092734] [PMID: 14963330]

[81]   Yu J, Zhang L. PUMA, a potent killer with or without p53. Oncogene 2008; 27 (Suppl. 1): S71-83.

[http://dx.doi.org/10.1038/onc.2009.45] [PMID: 19641508]

[82]   Chipuk JE, Moldoveanu T, Llambi F, Parsons MJ, Green DR. The BCL-2 family reunion. Mol Cell 2010; 37(3): 299-310.
        [http://dx.doi.org/10.1016/j.molcel.2010.01.025]

[83]   Letai A, Bassik MC, Walensky LD, Sorcinelli MD, Weiler S, Korsmeyer SJ. Distinct BH3 domains either sensitize or activate mitochondrial apoptosis, serving as prototype cancer therapeutics. Cancer Cell 2002; 2(3): 183-92.
        [http://dx.doi.org/10.1016/S1535-6108(02)00127-7] [PMID: 12242151]

[84]   Kuwana T, Bouchier-Hayes L, Chipuk JE, *et al.* BH3 domains of BH3-only proteins differentially regulate Bax-mediated mitochondrial membrane permeabilization both directly and indirectly. Mol Cell 2005; 17(4): 525-35.
        [http://dx.doi.org/10.1016/j.molcel.2005.02.003] [PMID: 15721256]

[85]   Hikisz P, Kiliańska ZM. PUMA, a critical mediator of cell death--one decade on from its discovery. Cell Mol Biol Lett 2012; 17(4): 646-69.
        [http://dx.doi.org/10.2478/s11658-012-0032-5] [PMID: 23001513]

[86]   Yu J, Zhang L. No PUMA, no death: implications for p53-dependent apoptosis. Cancer Cell 2003; 4(4): 248-9.
        [http://dx.doi.org/10.1016/S1535-6108(03)00249-6] [PMID: 14585351]

[87]   Ploner C, Kofler R, Villunger A. Noxa: at the tip of the balance between life and death. Oncogene 2008; 27 (Suppl. 1): S84-92.
        [http://dx.doi.org/10.1038/onc.2009.46] [PMID: 19641509]

[88]   Pronsato L, Milanesi L, Vasconsuelo A, La Colla A. Testosterone modulates FoxO3a and p53-related genes to protect C2C12 skeletal muscle cells against apoptosis. Steroids 2017; 124: 35-45.
        [http://dx.doi.org/10.1016/j.steroids.2017.05.012] [PMID: 28554727]

[89]   La Colla A, Vasconsuelo A, Boland R. Estradiol exerts antiapoptotic effects in skeletal myoblasts *via* mitochondrial PTP and MnSOD. J Endocrinol 2013; 216(3): 331-41.
        [http://dx.doi.org/10.1530/JOE-12-0486] [PMID: 23213199]

[90]   Ihrie RA, Reczek E, Horner JS, *et al.* Perp is a mediator of p53-dependent apoptosis in diverse cell types. Curr Biol 2003; 13: 1985e1990
        [http://dx.doi.org/10.1016/j.cub.2003.10.055]

[91]   Reczek EE, Flores ER, Tsay AS, Attardi LD, Jacks T. Multiple response elements and differential p53 binding control Perp expression during apoptosis. Mol Cancer Res 2003; 1: 1048e1057.

[92]   Pronsato L, Milanesi L. Effect of testosterone on the regulation of p53 and p66Shc during oxidative stress damage in C2C12 cells. Steroids 2016; 106: 41-54.
        [http://dx.doi.org/10.1016/j.steroids.2015.12.007] [PMID: 26703444]

[93]   Korsmeyer SJ. BCL-2 gene family and the regulation of programmed cell death. Cancer Res 1999; 59(7) (Suppl.): 1693s-700s.
        [PMID: 10197582]

[94]   Vaux DL, Cory S, Adams JM. Bcl-2 gene promotes haemopoietic cell survival and cooperates with c-myc to immortalize pre-B cells. Nature 1988; 335(6189): 440-2.
        [http://dx.doi.org/10.1038/335440a0] [PMID: 3262202]

[95]   Gross A, McDonnell JM, Korsmeyer SJ. BCL-2 family members and the mitochondria in apoptosis. Genes Dev 1999; 13(15): 1899-911.
        [http://dx.doi.org/10.1101/gad.13.15.1899] [PMID: 10444588]

[96]   Shamas-Din A, Kale J, Leber B, Andrews DW. Mechanisms of action of Bcl-2 family proteins. Cold Spring Harb Perspect Biol 2013; 5(4): a008714.
        [http://dx.doi.org/10.1101/cshperspect.a008714] [PMID: 23545417]

[97]   Seto M, Jaeger U, Hockett RD, *et al.* Alternative promoters and exons, somatic mutation and deregulation of the Bcl-2-Ig fusion gene in lymphoma. EMBO J 1988; 7(1): 123-31.
[http://dx.doi.org/10.1002/j.1460-2075.1988.tb02791.x] [PMID: 2834197]

[98]   Perillo B, Sasso A, Abbondanza C, Palumbo G. 17beta-estradiol inhibits apoptosis in MCF-7 cells, inducing bcl-2 expression *via* two estrogen-responsive elements present in the coding sequence. Mol Cell Biol 2000; 20(8): 2890-901.
[http://dx.doi.org/10.1128/MCB.20.8.2890-2901.2000] [PMID: 10733592]

[99]   Hyder SM, Nawaz Z, Chiappetta C, Yokoyama K, Stancel GM. The protooncogene c-jun contains an unusual estrogen-inducible enhancer within the coding sequence. J Biol Chem 1995; 270(15): 8506-13.
[http://dx.doi.org/10.1074/jbc.270.15.8506] [PMID: 7721748]

[100]   Strasser A, O'Connor L, Dixit VM. Apoptosis signaling. Annu Rev Biochem 2000; 69: 217-45.
[http://dx.doi.org/10.1146/annurev.biochem.69.1.217] [PMID: 10966458]

[101]   O'Connor L, Strasser A, O'Reilly LA, *et al.* Bim: a novel member of the Bcl-2 family that promotes apoptosis. EMBO J 1998; 17(2): 384-95.
[http://dx.doi.org/10.1093/emboj/17.2.384] [PMID: 9430630]

[102]   Chen L, Willis SN, Wei A, *et al.* Differential targeting of prosurvival Bcl-2 proteins by their BH3-only ligands allows complementary apoptotic function. Mol Cell 2005; 17(3): 393-403.
[http://dx.doi.org/10.1016/j.molcel.2004.12.030] [PMID: 15694340]

[103]   Liu B, Chen Y, St Clair DK. ROS and p53: a versatile partnership. Free Radic Biol Med 2008; 44(8): 1529-35.
[http://dx.doi.org/10.1016/j.freeradbiomed.2008.01.011] [PMID: 18275858]

[104]   Trinei M, Giorgio M, Cicalese A, *et al.* A p53-p66Shc signalling pathway controls intracellular redox status, levels of oxidation-damaged DNA and oxidative stress-induced apoptosis. Oncogene 2002; 21(24): 3872-8.
[http://dx.doi.org/10.1038/sj.onc.1205513] [PMID: 12032825]

[105]   Branco AF, Sampaio SF, Wieckowski MR, Sardão VA, Oliveira PJ. Mitochondrial disruption occurs downstream from β-adrenergic overactivation by isoproterenol in differentiated, but not undifferentiated H9c2 cardiomyoblasts: differential activation of stress and survival pathways. Int J Biochem Cell Biol 2013; 45(11): 2379-91.
[http://dx.doi.org/10.1016/j.biocel.2013.08.006] [PMID: 23958426]

[106]   Migliaccio E, Giorgio M, Mele S, *et al.* The p66shc adaptor protein controls oxidative stress response and life span in mammals. Nature 1999; 402(6759): 309-13.
[http://dx.doi.org/10.1038/46311] [PMID: 10580504]

[107]   Pelicci G, Lanfrancone L, Grignani F, *et al.* A novel transforming protein (SHC) with an SH2 domain is implicated in mitogenic signal transduction. Cell 1992; 70(1): 93-104.
[http://dx.doi.org/10.1016/0092-8674(92)90536-L] [PMID: 1623525]

[108]   Migliaccio E, Mele S, Salcini AE, *et al.* Opposite effects of the p52shc/p46shc and p66shc splicing isoforms on the EGF receptor-MAP kinase-fos signalling pathway. EMBO J 1997; 16(4): 706-16.
[http://dx.doi.org/10.1093/emboj/16.4.706] [PMID: 9049300]

[109]   Natalicchio A, De Stefano F, Perrini S, *et al.* Involvement of the p66Shc protein in glucose transport regulation in skeletal muscle myoblasts. Am J Physiol Endocrinol Metab 2009; 296(2): E228-37.
[http://dx.doi.org/10.1152/ajpendo.90347.2008] [PMID: 18957618]

[110]   Natalicchio A, Tortosa F, Perrini S, Laviola L, Giorgino F. p66Shc, a multifaceted protein linking Erk signalling, glucose metabolism, and oxidative stress. Arch Physiol Biochem 2011; 117(3): 116-24.
[http://dx.doi.org/10.3109/13813455.2011.562513] [PMID: 21506908]

[111]   Chen Z, Wang G, Zhai X, *et al.* Selective inhibition of protein kinase C β2 attenuates the adaptor P66 Shc-mediated intestinal ischemia-reperfusion injury. Cell Death Dis 2014; 5: e1164.

[http://dx.doi.org/10.1038/cddis.2014.131] [PMID: 24722289]

[112] Pinton P, Rimessi A, Marchi S, *et al.* Protein kinase C beta and prolyl isomerase 1 regulate mitochondrial effects of the life-span determinant p66Shc. Science 2007; 315(5812): 659-63.
[http://dx.doi.org/10.1126/science.1135380] [PMID: 17272725]

[113] Le S, Connors TJ, Maroney AC. c-Jun N-terminal kinase specifically phosphorylates p66ShcA at serine 36 in response to ultraviolet irradiation. J Biol Chem 2001; 276(51): 48332-6.
[http://dx.doi.org/10.1074/jbc.M106612200] [PMID: 11602589]

[114] Giorgio M, Migliaccio E, Orsini F, *et al.* Electron transfer between cytochrome c and p66Shc generates reactive oxygen species that trigger mitochondrial apoptosis. Cell 2005; 122(2): 221-33.
[http://dx.doi.org/10.1016/j.cell.2005.05.011] [PMID: 16051147]

[115] Li J, Kurokawa M. Regulation of MDM2 stability after DNA damage. J Cell Physiol 2015; 230(10): 2318-27.
[http://dx.doi.org/10.1002/jcp.24994] [PMID: 25808808]

[116] Meek DW. Mechanisms of switching on p53: a role for covalent modification? Oncogene 1999; 18(53): 7666-75.
[http://dx.doi.org/10.1038/sj.onc.1202951] [PMID: 10618706]

[117] Léger JG, Montpetit ML, Tenniswood MP. Characterization and cloning of androgen-repressed mRNAs from rat ventral prostate. Biochem Biophys Res Commun 1987; 147(1): 196-203.
[http://dx.doi.org/10.1016/S0006-291X(87)80106-7] [PMID: 3632663]

[118] Rosenberg ME, Silkensen J. Clusterin: physiologic and pathophysiologic considerations. Int J Biochem Cell Biol 1995; 27(7): 633-45.
[http://dx.doi.org/10.1016/1357-2725(95)00027-M] [PMID: 7648419]

[119] Wilson MR, Easterbrook-Smith SB. Clusterin is a secreted mammalian chaperone. Trends Biochem Sci 2000; 25(3): 95-8.
[http://dx.doi.org/10.1016/S0968-0004(99)01534-0] [PMID: 10694874]

[120] Trougakos IP, Gonos ES. Clusterin/apolipoprotein J in human aging and cancer. Int J Biochem Cell Biol 2002; 34(11): 1430-48.
[http://dx.doi.org/10.1016/S1357-2725(02)00041-9] [PMID: 12200037]

[121] Shannan B, Seifert M, Leskov K, *et al.* Challenge and promise: roles for clusterin in pathogenesis, progression and therapy of cancer. Cell Death Differ 2006; 13(1): 12-9.
[http://dx.doi.org/10.1038/sj.cdd.4401779] [PMID: 16179938]

[122] Buttyan R, Olsson CA, Pintar J, *et al.* Induction of the TRPM-2 gene in cells undergoing programmed death. Mol Cell Biol 1989; 9(8): 3473-81.
[http://dx.doi.org/10.1128/MCB.9.8.3473] [PMID: 2477686]

[123] Miyake H, Nelson C, Rennie PS, Gleave ME. Testosterone-repressed prostate message-2 is an antiapoptotic gene involved in progression to androgen independence in prostate cancer. Cancer Res 2000; 60(1): 170-6.
[PMID: 10646870]

[124] Zakeri Z, Curto M, Hoover D, *et al.* Developmental expression of the S35-S45/SGP-2/TRPM-2 gene in rat testis and epididymis. Mol Reprod Dev 1992; 33(4): 373-84.
[http://dx.doi.org/10.1002/mrd.1080330403] [PMID: 1472369]

[125] Bruchovsky N, Snoek R, Rennie PS, Akakura K, Goldenberg LS, Gleave M. Control of tumor progression by maintenance of apoptosis. Prostate Suppl 1996; 6: 13-21.
[http://dx.doi.org/10.1002/(SICI)1097-0045(1996)6+<13::AID-PROS4>3.0.CO;2-L]   [PMID: 8630223]

[126] Oliveira AS, Corbo M, Duigou G, Gabbai AA, Hays AP. Expression of a cell death marker (Clusterin) in muscle target fibers. Arq Neuropsiquiatr 1993; 51(3): 371-6.
[http://dx.doi.org/10.1590/S0004-282X1993000300014] [PMID: 8297243]

[127] Koltai T. Clusterin: a key player in cancer chemoresistance and its inhibition. OncoTargets Ther 2014; 7: 447-56.
[http://dx.doi.org/10.2147/OTT.S58622] [PMID: 24672247]

[128] Criswell T, Klokov D, Beman M, Lavik JP, Boothman DA. Repression of IR-inducible clusterin expression by the p53 tumor suppressor protein. Cancer Biol Ther 2003; 2(4): 372-80.
[http://dx.doi.org/10.4161/cbt.2.4.430] [PMID: 14508108]

[129] Boissonneault G. Evidence of apoptosis in the castration-induced atrophy of the rat levator ani muscle. Endocr Res 2001; 27(3): 317-28.
[http://dx.doi.org/10.1081/ERC-100106009] [PMID: 11678579]

[130] Arvill A, Adolfsson S, Ahren K. The use of the levator ani muscle *in vitro* in evaluating hormone action. Methods Enzymol 1975; 39: 94-101.
[http://dx.doi.org/10.1016/S0076-6879(75)39013-7] [PMID: 1152688]

[131] Tobin C, Joubert Y. Testosterone-induced development of the rat levator ani muscle. Dev Biol 1991; 146(1): 131-8.
[http://dx.doi.org/10.1016/0012-1606(91)90453-A] [PMID: 2060699]

[132] Tingus SJ, Carlsen RC. Effect of continuous infusion of an anabolic steroid on murine skeletal muscle. Med Sci Sports Exerc 1993; 25(4): 485-94.
[PMID: 8479303]

[133] Souccar C, Lapa AJ, Ribeiro do Valle J. Influence of castration on the electrical excitability and contraction properties of the rat levator ani muscle. Exp Neurol 1982; 75(3): 576-88.
[http://dx.doi.org/10.1016/0014-4886(82)90026-7] [PMID: 7060688]

[134] O'Rourke JR, Georges SA, Seay HR, *et al.* Essential role for Dicer during skeletal muscle development. Dev Biol 2007; 311 359e368
[http://dx.doi.org/10.1016/j.ydbio.2007.08.032]

[135] Callis TE, Chen JF, Wang DZ. MicroRNAs in skeletal and cardiac muscle development. DNA Cell Biol 2007; 26: 219e225.
[http://dx.doi.org/10.1089/dna.2006.0556]

[136] Chen JF, Tao Y, Li J, *et al.* microRNA-1 and microRNA-206 regulate skeletal muscle satellite cell proliferation and differentiation by repressing Pax7. J Cell Biol 2010; 190(5): 867-79.
[http://dx.doi.org/10.1083/jcb.200911036] [PMID: 20819939]

[137] McCarthy JJ, Esser KA. MicroRNA-1 and microRNA-133a expression are decreased during skeletal muscle hypertrophy. J Appl Physiol 2007; 102(1): 306-13.
[http://dx.doi.org/10.1152/japplphysiol.00932.2006] [PMID: 17008435]

[138] McCarthy JJ. MicroRNA-206: the skeletal muscle-specific myomiR. Biochim Biophys Acta 2008; 1779(11): 682-91.
[http://dx.doi.org/10.1016/j.bbagrm.2008.03.001] [PMID: 18381085]

[139] Townley-Tilson WH, Callis TE, Wang D. MicroRNAs 1, 133, and 206: critical factors of skeletal and cardiac muscle development, function, and disease. Int J Biochem Cell Biol 2010; 42(8): 1252-5.
[http://dx.doi.org/10.1016/j.biocel.2009.03.002] [PMID: 20619221]

[140] Yuasa K, Hagiwara Y, Ando M, Nakamura A, Takeda S, Hijikata T. MicroRNA-206 is highly expressed in newly formed muscle fibers: implications regarding potential for muscle regeneration and maturation in muscular dystrophy. Cell Struct Funct 2008; 33(2): 163-9.
[http://dx.doi.org/10.1247/csf.08022] [PMID: 18827405]

[141] Cheung TH, Quach NL, Charville GW, *et al.* Maintenance of muscle stem-cell quiescence by microRNA-489. Nature 2012; 482(7386): 524-8.
[http://dx.doi.org/10.1038/nature10834] [PMID: 22358842]

[142] Crist CG, Montarras D, Buckingham M. Muscle satellite cells are primed for myogenesis but maintain

quiescence with sequestration of Myf5 mRNA targeted by microRNA-31 in mRNP granules. Cell Stem Cell 2012; 11(1): 118-26.
[http://dx.doi.org/10.1016/j.stem.2012.03.011] [PMID: 22770245]

[143] Shenoy A, Blelloch RH. Regulation of microRNA function in somatic stem cell proliferation and differentiation. Nat Rev Mol Cell Biol 2014; 15(9): 565-76.
[http://dx.doi.org/10.1038/nrm3854] [PMID: 25118717]

[144] Huang Z, Chen X, Yu B, He J, Chen D. MicroRNA-27a promotes myoblast proliferation by targeting myostatin. Biochem Biophys Res Commun 2012; 423(2): 265-9.
[http://dx.doi.org/10.1016/j.bbrc.2012.05.106] [PMID: 22640741]

[145] Gagan J, Dey BK, Layer R, Yan Z, Dutta A. MicroRNA-378 targets the myogenic repressor MyoR during myoblast differentiation. J Biol Chem 2011; 286(22): 19431-8.
[http://dx.doi.org/10.1074/jbc.M111.219006] [PMID: 21471220]

[146] Liu J, Luo XJ, Xiong AW, *et al.* MicroRNA-214 promotes myogenic differentiation by facilitating exit from mitosis *via* down-regulation of proto-oncogene N-ras. J Biol Chem 2010; 285(34): 26599-607.
[http://dx.doi.org/10.1074/jbc.M110.115824] [PMID: 20534588]

[147] Gambardella S, Rinaldi F, Lepore SM, *et al.* Overexpression of microRNA-206 in the skeletal muscle from myotonic dystrophy type 1 patients. J Transl Med 2010; 8: 48.
[http://dx.doi.org/10.1186/1479-5876-8-48] [PMID: 20487562]

[148] Goljanek-Whysall K, Sweetman D, Münsterberg AE. microRNAs in skeletal muscle differentiation and disease. Clin Sci (Lond) 2012; 123(11): 611-25.
[http://dx.doi.org/10.1042/CS20110634] [PMID: 22888971]

[149] Kim JY, Park YK, Lee KP, *et al.* Genome-wide profiling of the microRNA-mRNA regulatory network in skeletal muscle with aging. Aging (Albany NY) 2014; 6(7): 524-44.
[http://dx.doi.org/10.18632/aging.100677] [PMID: 25063768]

[150] Rosales XQ, Malik V, Sneh A, *et al.* Impaired regeneration in LGMD2A supported by increased PAX7-positive satellite cell content and muscle-specific microrna dysregulation. Muscle Nerve 2013; 47(5): 731-9.
[http://dx.doi.org/10.1002/mus.23669] [PMID: 23553538]

[151] Hudson MB, Woodworth-Hobbs ME, Zheng B, *et al.* miR-23a is decreased during muscle atrophy by a mechanism that includes calcineurin signaling and exosome-mediated export. Am J Physiol Cell Physiol 2014; 306(6): C551-8.
[http://dx.doi.org/10.1152/ajpcell.00266.2013] [PMID: 24336651]

[152] Wada S, Kato Y, Okutsu M, *et al.* Translational suppression of atrophic regulators by microRNA-23a integrates resistance to skeletal muscle atrophy. J Biol Chem 2011; 286(44): 38456-65.
[http://dx.doi.org/10.1074/jbc.M111.271270] [PMID: 21926429]

[153] Hitachi K, Tsuchida K. Role of microRNAs in skeletal muscle hypertrophy. Front Physiol 2014; 4: 408.
[http://dx.doi.org/10.3389/fphys.2013.00408] [PMID: 24474938]

[154] Javed R, Jing L, Yang J, Li X, Cao J, Zhao S. miRNA transcriptome of hypertrophic skeletal muscle with overexpressed myostatin propeptide. BioMed Res Int 2014; 2014: 328935.
[http://dx.doi.org/10.1155/2014/328935] [PMID: 25147795]

[155] Luo W, Nie Q, Zhang X. MicroRNAs involved in skeletal muscle differentiation. J Genet Genomics 2013; 40(3): 107-16.
[http://dx.doi.org/10.1016/j.jgg.2013.02.002] [PMID: 23522383]

[156] Ge Y, Sun Y, Chen J. IGF-II is regulated by microRNA-125b in skeletal myogenesis. J Cell Biol 2011; 192(1): 69-81.
[http://dx.doi.org/10.1083/jcb.201007165] [PMID: 21200031]

[157] Cleveland BM, Weber GM. Effects of sex steroids on indices of protein turnover in rainbow trout

(Oncorhynchusmykiss) white muscle. Gen Comp Endocrinol 2011; 174(2): 132-42.
[http://dx.doi.org/10.1016/j.ygcen.2011.08.011] [PMID: 21878334]

[158]   Cleveland BM, Weber GM. Effects of sex steroids on expression of genes regulating growth-related mechanisms in rainbow trout (Oncorhynchus mykiss). Gen Comp Endocrinol 2015; 216: 103-15.
[http://dx.doi.org/10.1016/j.ygcen.2014.11.018] [PMID: 25482545]

[159]   Salem M, Kenney PB, Rexroad CE, Yao J. Molecular characterization of muscle atrophy and proteolysis associated with spawning in rainbow trout. Comp Biochem Physiol Part D Genomics Proteomics 2006; 1(2): 227-37.
[http://dx.doi.org/10.1016/j.cbd.2005.12.003] [PMID: 20483254]

[160]   Koganti PP, Wang J, Cleveland B, Ma H, Weber GM, Yao J. Estradiol regulates expression of miRNAs associated with myogenesis in rainbow trout. Mol Cell Endocrinol 2017; 443: 1-14.
[http://dx.doi.org/10.1016/j.mce.2016.12.014] [PMID: 28011237]

[161]   Tait SWG, Green DR. Mitochondria and cell death: outer membrane permeabilization and beyond. Nat Rev Mol Cell Biol 2010; 11(9): 621-32.
[http://dx.doi.org/10.1038/nrm2952] [PMID: 20683470]

CHAPTER 5

Antiapoptotic Effects of Estrogens and Androgens

Lorena M. Milanesi[*]

Instituto de Ciencias Biológicas y Biomédicas de Sur (INBIOSUR)-UNS-CONICET, Bahía Blanca, Argentina

Abstract: As it was mentioned before, estrogens and androgens acts on different tissues, also estrogens and androgens receptors are ubiquitously expressed and have shown not only nuclear, but also non-classical intracellular sites like plasma membrane, mitochondria, golgi and endoplasmic reticulum, making increasing this properties a more complex function to the classical roles of estrogens and androgens (regulation of gene expression). E2 and T can trigger different pathways by a non-genomic mechanism through proteins that have the ability to interact with the steroid hormones (structural similar or different from known steroid receptors). So the hormones can regulate apoptotic events through those different signaling pathways.

In mitochondria, a control point of apoptosis, it was demonstrated not only the presence of ER and AR but also an steroid protective action against different injuries that results in antiapoptotic effect. Here we summarize the molecular events, modulated by E2 and/or T in several tissues, during programmed cell death.

Keywords: Apoptosis, C2C12, Estradiol, Proliferation, Testosterone.

INTRODUCTION

Experimental data have shown that estrogens and androgens control apoptosis using different pathways [reviewed in 1]. In these events, the sex hormones acts through steroid binding proteins with classical or non classical locations.

Estrogens modulate equilibrium between cell survival, proliferation and death.

Though, how estrogens change balance in direction of survival or apoptosis it is not fully explained.

Protective Actions of E2 and T

In ER-positive tumoral cells MCF-7, estrogens induce transition of G1 to S phase.

---

[*] **Corresponding author Lorena M Milanesi:** Instituto de Ciencias Biológicas y Biomédicas de Sur (INBIOSUR)-UNS-CONICET, Bahía Blanca, Argentina; E-mail: milanesi@criba.edu.ar

This action is connected with the expression of c-myc, and the activity of cyclin D1, cyclin-dependent kinase (CDK) and retinoblastoma protein [2].

Though, under some precise circumstances estrogens induce apoptosis in other breast cancer cells lines, contrasting to the antiapoptotic function.

This unusual comportment it was also observed in breast cancer cells which have been E2- deprived (LTED) or treated thoroughly with anti-estrogens [3].

Pre-conditions of prolonged E2 reduction or exhaustive treatment using anti-estrogens are basics to cause apoptosis and could clarify the dual role observed: survival, proliferation or programmed cell death.

The author proposes that regardless of the fact that the estrogen receptors still regulates gene expression, also can activate the Fas apoptotic cascade or otherwise has a direct action on mitochondria through downregulation antiapoptotic members of the bcl-2 family, leading this into apoptosis [3].

In C2C12 myoblats cell line, T and E2, in nanomolar dose avoid H_2O_2-induced apoptosis [4, 5]. Typical changes induced by H_2O_2 such as cytoskeleton alteration, mitochondrial clustering/dysfunction, cytochrome c release and nuclear fragmentation, are reduced when C2C12 cells are treated with the steroid hormones.

Various molecular actions that occur during anti-apoptotic actions of androgens on C2C12 cell line have also been recognized. Short treatments with hydrogen peroxide, induce a rapid cellular defense response involving ERK2, Akt and Bad phosphorylation and an increase of HSP70 protein levels. At longer periods of treatment with the apoptotic agent, dephosphorylation of these proteins, cytochrome c release, PARP cleavage and DNA degradation arises.

Specifically for T, Previous hormone treatment generate Bad inactivation (phosphorylation), translocation of HSP90 to mitocondria, increase in c actin levels and reduction in Bax levels [5, 6] (Fig. **1**).

Also in this cell line T diminishes p-p53 and keeps the inactive state of FoxO3a transcription factor, induced by H_2O_2 [7].

Increased gene expression of the proapoptotic genes Puma, PERP, Bim and MDM2 and the downregulation of Bcl-2 are observed during H_2O_2 treatment at diverse periods of time during apoptosis.

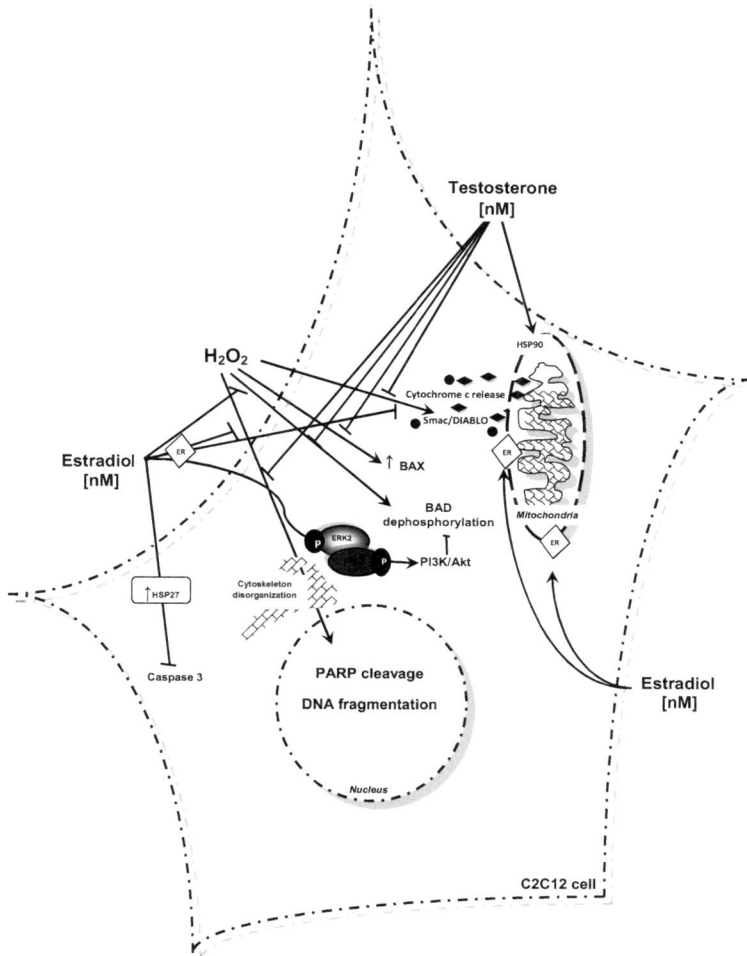

Fig. (1). Events implicated in the survival action of T and E2 in C2C12. Both sex steroids interact with apoptotic mediators and trigger diverse intracellular signaling cascade to inhibit H_2O_2-induced apoptosis. E2 can interact with estradiol binding proteins/receptors (ER) confined in plasmatic membrane or mitochondria that encourages activation of the PI3K/Akt/P-BAD survival pathway, ERK, p38 MAPK, HSP27 and inhibition of caspase-3 function. E2 abolishes mitochondrial membrane injury, and thus Smac/DIABLO and cytochrome c release, induced by H_2O_2. Comparable to E2, T is capable to protect the mitochondria. T abolishes H_2O_2-induced bax expression. Steroids act to avoid DNA degradation and cytoskeleton disruption [Extracted from 1].

T is capable to control and neutralize this effects supporting that the antiapoptotic effect of the hormone is also exerting at the transcriptional level. Besides, ERK and JNK kinases have been established to be associated to FoxO3a phosphorylation and thus its intracellular localization. E2, at physiological concentrations, could reduce apoptosis damage in C2C12 through estrogen receptors concerning MAPKs, HSP27, PI3K/Akt pathway, which in turn deactivates, members of the Bcl-2 family, PKCδ, JNK and p66Shc, exerting a positive action in the mitochondria [8].

E2 also modulates, as T, p53 and FoxO transcription factors and in consequence their target genes Bcl-2, Bim, Puma, PERP and MDM2, without disturbing Noxa expression levels, supporting the idea that FoxO and p53 synchronize apoptosis in C2C12 cell line [4 - 10].

In the nervous system, E2 and T have a clearly protective effect against apoptosis.

Physiological levels of the steroid hormones are neuroprotective both *in vivo* and *in vitro* and specifically for E2 the protective action is mainly due through mitochondria [11].

For E2 the final result in the nervous system depends on the subtype of the receptor: cells which express higher levels of ERβ undergo apoptosis, whereas cells expressing higher levels of ERα are protected from apoptosis [12].

Experimental approaches indicated that ERβ mediates apoptosis by a mechanism that needs FasL [12]. Nevertheless, in *in vivo* assays using βERKO mice, established that ERβ is essential for neuronal subsistence [13].

Finally, the response to the hormone could be not only hormone but also sex specific. Thus, in T47D breast cancer cells contrary effects were observed for E2 and T non-permeable ligands. E2 protected cells from serum-deprivation-induced death, while T triggered apoptosis in serum supplemented T47D cells.

The mechanism for T in neuroprotection is fewer understand but the participation of androgen receptor is well established by the employed of the AR antagonist flutamide. Using an *in vitro/ex vivo* model, it was demonstrated that cerebellar granule cells (CGC) from neonatal rats treated with T are less susceptible to 50 mM H_2O_2 treatment [13, 14]. They verified a mechanism concerning an upregulation of the cellular antioxidant defenses, specifically an increased activity of catalase and superoxide dismutase.

In cultured hippocampal neurons from rats and in PC12 cells stably transfected with the androgen receptor, T and dihydrotestosterone (DHT) protected against apoptosis induced by β-amyloid peptides (Aβ) [15]. 10 nM of T or DHT promptly increased ERK-1 and ERK-2, Rsk-1, and Bad phosphorylation.

This effect arises only in cells expressing functional AR. Neuroprotective androgen signaling does not seem to implicate cell surface-associated receptors as the use of androgen-impermeable ligands -BSA conjugates failed to activate MAPK/ERK pathways.

In β cells of pancreas, apoptosis is responsible for the progression of insulin-dependent diabetes mellitus in the streptozotocin (STZ) rat model. In this experimental model, Morimoto *et al.* [16] established that castrated rats showed higher percentages of apoptotic β cells than intact males and castrated, T-substituted males (castrated rats were replaced with T enanthate). The reduction in apoptotic pancreatic cells induced by T was inhibited by flutamide, showing a possible participation of the AR in the beneficial role of the male hormone. Furthermore, the precise effect of the steroid hormone is linked to the expression and activity of antioxidant enzyme in pancreatic β cells [17].

This protective effect is again sex specific. Cytoprotective effect was detected in T but not in progesterone or E2 treated male rats, and it was seen in male but not in female animals.

It is recognized that at physiological levels, T is involved in neuronal differentiation, neuroprotection, survival and development [18, 19]. However, androgens like T could also apply apoptotic effects in cells of neuroblastoma (SHSY5Y) at supraphysiological concentrations (micro molar range) starting the apoptotic pathway, and is abrogated in the presence of either inhibitors of caspases or the inositol 1,4,5-trisphosphate receptor (InsP3R)-mediated calcium release. It has been established that T and 5 alpha-DHT induce programmed cell death in cultured dermal papilla cells (DPC) in a dose-dependent and time-related mode with a decline in Bcl-2 protein expression, a rise in the Bax/Bcl-2 ratio and caspase-8 induction.

Another investigation showing the role of androgen in apoptosis and its influence on cancer treatment has been made in the Caco-2 cell line (from human epithelial colorectal adenocarcinoma). In this experimental system, by means of fluorescent non-membrane-permeable AR ligands (T-HAS-FITC), Gu *et al.* [20] observed not only the existence of membrane ARs but also that the prompt with T conjugates (T-HAS) induced fast cytoskeleton reorganization and apoptotic reactions through activation of the caspase-3, even in the presence of anti-androgens.

These properties were particular for T and its conjugates, since other steroids, for instance E2 did not display any pro-apoptotic effects. These facts add important portion of evidence to the role of membrane androgen receptors during apoptosis. Therefore, their activation by steroid albumin conjugates prompts strong pro-apoptotic reactions related to caspase induction and cytoskeletal reorganizations.

When conjugated ligands are employed we have to be careful to investigate functions of membrane receptors. These ligands are not too much stable and the experiential result is not merely due to the conjugate ligand that activates a specific membrane receptor, but also to portions of unconjugated compound (hormone without the BSA fraction) that could go through the plasma membrane and interact with intracellular receptors.

A resolution to this problem could be to make a totally inhibition of the intracellular receptors and then analyze, under these conditions, the role of membrane hormone receptors.

CONCLUDING REMARKS

E2 and T regulate several other functions in adding to reproduction, in a diversity of mammal tissues expressing estrogen and androgens receptors with or without typical intracellular locations or other steroid receptors like G protein-coupled receptors. Programmed cell death is a significant process underlying the effects of both sex hormones on *target* tissues. E2 and T control apoptosis mainly through receptors placed in mitochondria, plasma membrane and endoplasmic reticulum. Binding of their ligands triggers the MAPK and PI3K/Akt pathways and induces antiapoptotic members of the Bcl-2 group, altogether leading to survival. However, in some cellular lines and with specific hormone doses, estrogens can also abrogates the PI3K/Akt cascade and then induce apoptosis. Male hormone can also trigger apoptotic effects through similar molecular mechanisms. Hence, this studies could clarify the basis by which the sex steroids upregulate or downregulate these pathways and may help to reveal novel therapeutic targets in pathologies linked to deregulation of apoptosis.

CONSENT FOR PUBLICATION

Not applicable.

CONFLICT OF INTEREST

The authors confirm that this chapter contents have no conflict of interest.

ACKNOWLEDGEMENTS

National University of the South Argentina and National Research Council of Argentina (CONICET).

REFERENCES

[1]   Vasconsuelo A, Pronsato L, Ronda AC, Boland R, Milanesi L. Role of 17β-estradiol and testosterone in apoptosis. Steroids 2011; 76(12): 1223-31.
[http://dx.doi.org/10.1016/j.steroids.2011.08.001] [PMID: 21855557]

[2] Altucci L, Addeo R, Cicatiello L, *et al.* 17beta-Estradiol induces cyclin D1 gene transcription, p36D1-p34cdk4 complex activation and p105Rb phosphorylation during mitogenic stimulation of G(1)-arrested human breast cancer cells. Oncogene 1996; 12(11): 2315-24.
[PMID: 8649771]

[3] Jordan VC. The 38th David A. Karnofsky lecture: the paradoxical actions of estrogen in breast cancer--survival or death? J Clin Oncol 2008; 26(18): 3073-82.
[http://dx.doi.org/10.1200/JCO.2008.17.5190] [PMID: 18519949]

[4] Vasconsuelo A, Milanesi L, Boland R. 17β-estradiol abrogates apoptosis in murine skeletal muscle cells through estrogen receptors: role of the phosphatidylinositol 3-kinase/Akt pathway. J Endocrinol 2008; 196(2): 385-97.
[http://dx.doi.org/10.1677/JOE-07-0250] [PMID: 18252962]

[5] Pronsato L, Ronda AC, Milanesi L, Vasconsuelo A, Boland R. Protective role of 17 -estradiol and testosterone in apoptosis of skeletal muscle. Actual Osteol 2010; 2: 45-8.

[6] Jordan VC. The 38th David A. Karnofsky lecture: the paradoxical actions of estrogen in breast cancer—survival or death?. J Clin Oncol 2012.

[7] Pronsato L, Milanesi L, Vasconsuelo A, La Colla A. Testosterone modulates FoxO3a and p53-related genes to protect C2C12 skeletal muscle cells against apoptosis. Steroids 2017; 124: 35-45.
[http://dx.doi.org/10.1016/j.steroids.2017.05.012] [PMID: 28554727]

[8] La Colla A, Vasconsuelo A, Milanesi L, Pronsato L. 17β-estradiol protects skeletal myoblasts from apoptosis through p53, bcl-2 and foxo families. J Cell Biochem 2017; 118(1): 104-15.
[http://dx.doi.org/10.1002/jcb.25616] [PMID: 27249370]

[9] Ronda AC, Vasconsuelo A, Boland R. Extracellular-regulated kinase and p38 mitogen-activated protein kinases are involved in the antiapoptotic action of 17beta-estradiol in skeletal muscle cells. J Endocrinol 2010; 206(2): 235-46.
[http://dx.doi.org/10.1677/JOE-09-0429] [PMID: 20488946]

[10] Vasconsuelo A, Milanesi L, Boland R. Participation of HSP27 in the antiapoptotic action of 17beta-estradiol in skeletal muscle cells. Cell Stress Chaperones 2010; 15(2): 183-92.
[http://dx.doi.org/10.1007/s12192-009-0132-y] [PMID: 19621276]

[11] Ahlbom E, Grandison L, Bonfoco E, Zhivotovsky B, Ceccatelli S. Androgen treatment of neonatal rats decreases susceptibility of cerebellar granule neurons to oxidative stress *in vitro*. Eur J Neurosci 1999; 11(4): 1285-91.
[http://dx.doi.org/10.1046/j.1460-9568.1999.00529.x] [PMID: 10103123]

[12] Sánchez-Más J, Turpín MC, Lax A, Ruipérez JA, Valdés Chávarri M, Pascual-Figal DA. Differential actions of eplerenone and spironolactone on the protective effect of testosterone against cardiomyocyte apoptosis *in vitro*. Rev Esp Cardiol 2010; 63(7): 779-87.
[http://dx.doi.org/10.1016/S0300-8932(10)70180-9] [PMID: 20609311]

[13] Nguyen TV, Yao M, Pike CJ. Androgens activate mitogen-activated protein kinase signaling: role in neuroprotection. J Neurochem 2005; 94(6): 1639-51.
[http://dx.doi.org/10.1111/j.1471-4159.2005.03318.x] [PMID: 16011741]

[14] Ahlbom E, Prins GS, Ceccatelli S. Testosterone protects cerebellar granule cells from oxidative stress-induced cell death through a receptor mediated mechanism. Brain Res 2001; 892(2): 255-62.
[http://dx.doi.org/10.1016/S0006-8993(00)03155-3] [PMID: 11172772]

[15] Kokontis JM, Hay N, Liao S. Progression of LNCaP prostate tumor cells during androgen deprivation: hormone-independent growth, repression of proliferation by androgen, and role for p27Kip1 in androgen-induced cell cycle arrest. Mol Endocrinol 1998; 12(7): 941-53.
[http://dx.doi.org/10.1210/mend.12.7.0136] [PMID: 9658399]

[16] Rubinow DR, Schmidt PJ. Androgens, brain, and behavior. Am J Psychiatry 1996; 153(8): 974-84.
[http://dx.doi.org/10.1176/ajp.153.8.974] [PMID: 8678193]

[17]   Estrada M, Varshney A, Ehrlich BE. Elevated testosterone induces apoptosis in neuronal cells. J Biol Chem 2006; 281(35): 25492-501.
[http://dx.doi.org/10.1074/jbc.M603193200] [PMID: 16803879]

[18]   Morimoto S, Mendoza-Rodríguez CA, Hiriart M, Larrieta ME, Vital P, Cerbón MA. Protective effect of testosterone on early apoptotic damage induced by streptozotocin in rat pancreas. J Endocrinol 2005; 187(2): 217-24.
[http://dx.doi.org/10.1677/joe.1.06357] [PMID: 16293769]

[19]   Palomar-Morales M, Morimoto S, Mendoza-Rodríguez CA, Cerbón MA. The protective effect of testosterone on streptozotocin-induced apoptosis in beta cells is sex specific. Pancreas 2010; 39(2): 193-200.
[http://dx.doi.org/10.1097/MPA.0b013e3181c156d9] [PMID: 20093993]

[20]   Gu S, Papadopoulou N, Gehring EM, *et al.* Functional membrane androgen receptors in colon tumors trigger pro-apoptotic responses *in vitro* and reduce drastically tumor incidence *in vivo*. Mol Cancer 2009; 8: 114.
[http://dx.doi.org/10.1186/1476-4598-8-114] [PMID: 19948074]

CHAPTER 6

Role Of Sex Hormones In Cytoskeletal Structure: Implications In Cellular Lifespan

Andrea Vasconsuelo[*]

Instituto de Ciencias Biológicas y Biomédicas del Sur (INBIOSUR), Universidad Nacional del Sur- CONICET, Bahía Blanca, Argentina

Abstract: The cytoskeleton is composed of intracellular structures that maintain cell shape, interconnect organelles to each other, often attached to the cell membrane and is involved in signaling pathways. Sex steroids are effective regulators of cell morphology and recent evidence indicates that it is obtained through the regulation of the actin cytoskeleton. Intriguingly, evidence implicates the actin cytoskeleton as both a sensor and mediator of apoptosis. Since it has been shown that sex hormones affect/regulate the cellular cytoskeleton, the aim of the present chapter is to discuss the molecular mechanism and targets, related to cytoskeleton, where sex steroids act and modulate the apoptotic process and in consequence the cellular lifespan.

Keywords: Actin, Cytoskeleton, Signalling Pathways.

INTRODUCTION

The term "cytoskeleton" defines a system of intracellular structures that maintain cell shape, interconnect organelles to each other, and often attach to the cell membrane. This system consists of several filamentous networks that extend from the plasma membrane to the nuclear membrane and even the interior of the nucleus. The cytoskeleton also plays a role in anchoring the cell to its neighbors and to the extracellular matrix by specialized cell junctions that span the plasma membrane. The cytoskeleton of eukaryotic cells is made up of filamentous proteins, all cytoskeletons consist of three major classes of elements that differ in size and in protein composition. Microtubules are the largest type of filament, with a diameter of about 25 nanometers (nm), containing primarily α- and β-tubulin [1]. Actin filaments, called microfilaments are the smallest type, with a diameter of only about 6-8 nm, and they are made up of a protein called actin.

[*] **Corresponding author Andrea A. Vasconsuelo:** Instituto de Ciencias Biológicas y Biomédicas del Sur (INBIOSUR), Universidad Nacional del Sur- CONICET, Bahía Blanca, Argentina; E-mail: avascon@criba.edu.ar

Intermediate filaments, as their name suggests, are mid-sized, with a diameter of about 10 nm. Unlike actin filaments and microtubules, intermediate filaments are built from a number of different subunit proteins (desmin, vimentin, keratins, *etc.*, depending on the cell type).

In skeletal muscle, the cytoskeleton may also be subdivided into the extra-sarcomeric (consists mainly of intermediate filaments), the intra-sarcomeric (consists mainly of titin and nebulin) and the subsarcolemmal (include proteins such as vinculin, spectrin, dystrophin, transmembrane integrins, ankyrin, α-actinin and desmin) cytoskeleton, depending on its relation with sarcolemma [reviews in 2]. Titin and nebulin, are two large filamentous proteins that are longitudinally arranged within the sarcomeres. Titin extends from the Z-disc to the M-band, and nebulin from the Z-disc along the length of the actin filaments [reviews in 2]. The protein of subsarcolemmal cytoskeleton, indirectly connects the most peripheral myofibrils with the extracellular matrix in specialized sarcolemmal domains, the costameres [3]. These cytoskeletal structures interact together as a dynamic network contributing to increase tensile strength, allow cell motility, maintain plasma integrity, participate in cell division and provide a network for cellular transport among others [4]. In addition, mounting evidence indicates that the cytoskeleton is involved in apoptosis, and that actin may be an early modulator of apoptotic commitment [5]. Previously we described characteristic morphological changes during apoptosis, to achieve such changes, apoptotic cells make profound cytoskeleton reorganizations, and caspase-mediated digestion of cytoskeleton proteins ensures the proper dismantlement of the dying cell during this process [6]. These cytoskeletal rearrangements have been mainly attributed to actino-myosin ring contraction, while microtubule and intermediate filaments are depolymerized at early stages of apoptosis. Nevertheless, microtubules also are involved in the execution phase of apoptosis forming an apoptotic microtubule network which maintains plasma membrane integrity and cell morphology during the execution phase of apoptosis [7]. The maintenance of this microtubule structure depends on high cellular ATP levels, so that microtubules undergo disorganization in late apoptosis, when mitochondria suffer extensive depolarization [8].

A large amount of literature supports that sex steroid hormones are fundamental modulators of cell morphology, through the regulation of the actin cytoskeleton [9]. Then, in view of the evidence indicating that cytoskeletal plays an important role triggering and during apoptosis and that it has been shown that sex hormones affect/regulate the cellular cytoskeleton, we can consider this filamentous structure as another point where sex steroids act and modulate the apoptotic process and in consequence the cellular lifespan.

Cytoskeleton Regulation by Sex Steroids: Implications in Apoptosis

The cytoskeleton has been demonstrated as essential during apoptosis with dramatic changes in actin filament organization associated with different stages of apoptosis [10]. The most logical data supporting links between actin and apoptosis come from investigations with drugs that affect actin turnover. The use of drugs to manipulate actin dynamics has been a valuable tool to investigate associations between actin and apoptosis signaling pathways. Even though the drugs used in these studies are widely accepted to influence the dynamic of the actin cytoskeleton, it is worth emphasizing the risk that uncharacterized actin-independent effects may also arise and, therefore, the results must be carefully analyzed taking into account all possible effect.

Actin is a ~42 kDa globular protein (G-actin) that reversibly polymerizes to form filaments (F-actin). In muscle cells, actin is a central component of the sarcomere and interacts with myosin filaments to allow force generation essential during muscle contraction [11]. The G-actin is spread evenly between cytoplasm and nucleoplasm; the distribution of actin filaments depends on the cell type, its role, and the cell cycle phase in which it is located [Review in 12]. Reorganization of the actin structure and the transition between its forms play a significant role in many of the most important cellular processes like cytokinesis, cell differentiation, and death. Recent studies have indicated that changes in the dynamic state of actin can change cell fate. In addition, increased turnover of filamentous (F)-actin can promote cell longevity, whereas reduced actin turnover looks to trigger apoptosis by mitochondria pathway [13]. The link between apoptosis eliciting and cytoskeleton dynamics is not restricted to scenarios in which F-actin structures are stabilized, as there is also evidence for a similar effect in some animal cells when actin is depolymerized. For example, in T-cells, actin depolymerization resulted in high caspase-3 activity [14].

The continuous assembly and disassembly process of actin filaments and the functions of the cytoskeleton are tightly regulated by the action of a plethora of proteins termed actin- binding proteins (ABPs), many of which are conserved from yeast to humans [15]. These actin-binding proteins undertake a range of functions that allows the actin cytoskeleton to constantly adapt to a fluctuating environment. ABPs are essential in regulating typical events of apoptotic processes such as cell rounding, membrane blebbing, caspase activation and mitochondrial membrane permeabilization [Review in 16]. ABPs known to play a role in apoptosis in animal cells include gelsolin, cofilin/ADF (actin-depolymerization factor), coronin and β-thymosins [Review in 17]. Interesting studies indicated that ABPs are prone to estrogenic regulation [18]. Gelsolin is a key regulator of cell survival in mammalian cells, working at multiple points within the regulation of apoptosis pathways. For example, it has been suggested that gelsolin is able to regulate VDACs, mitochondrial membrane pores responsible for the release of pro-apoptotic factors and that are central for maintaining mitochondrial membrane potential [Review in 17]. However, since gelsolin also is identified as a substrate for caspase-3, its dual roles in promoting apoptosis and protecting cells from apoptosis are proposed [19, 20]. The N-terminus of gelsolin is sensitive to phosphatidylinositol bisphosphate (PIP$_2$) that controls the uncapping of gelsolin, and the C-terminus of gelsolin is sensitive to Ca^{2+}, The C-terminus is also the regulatory domain that controls the severing activity of the N-terminus. The linkage between the N- and C-terminus of gelsolin is the site of cleavage for caspase-3. As soon as gelsolin is cleaved, Ca^{2+} does not control the severing activity of the N-terminus and consequently collapse of the cytoskeleton destructs cell integrity that usually accompanies the apoptosis [Review in 21]. Remarkable, it has been demonstrated that gelsolin acts as coregulator of androgen receptor (AR). Two binding sites, identified in the C-terminal ligand-binding domain and the central DNA-binding domain of the AR, can bind gelsolin. Gelsolin, in turn, increases the transcriptional activity of the AR in the presence of agonist. During the agonist-induced translocation of androgen receptor into the nucleus, gelsolin migrates to the perinuclear region. This localization is dependent on the presence of AR. The up regulation of gelsolin by androgen depletion suggests that gelsolin may be under negative feedback control *via* androgen [Review in 22]. Undoubtedly, sexual steroids play a key role regulating apoptosis in skeletal muscle by means of a multiple targets as mitochondria, nuclei, members of apoptotic cascades, ABPs, kinases and phosphatases, among others, which form an intricate net of signaling rigorously controlled, which is attached on the cytoskeleton.

C2C12 cells exposed to H$_2$O$_2$ resembling a senescent phenotype similar to aged skeletal muscle [23], since studies suggested that acute exposure to H$_2$O$_2$ triggers the appearance of a senescent-like phenotype, such as senescent-like morphology, shorter lifespan, growth arrest, and decrease proliferation capacity. This represents a useful tool to understand the molecular mechanisms implicated in sarcopenia of the older persons. Under these conditions was observed an important disorganization of cytoskeleton in C2C12 cells and in culture primary of mouse skeletal muscle cells, which in presence of estradiol or testosterone was abolished (Fig. **1**) [24, 25].

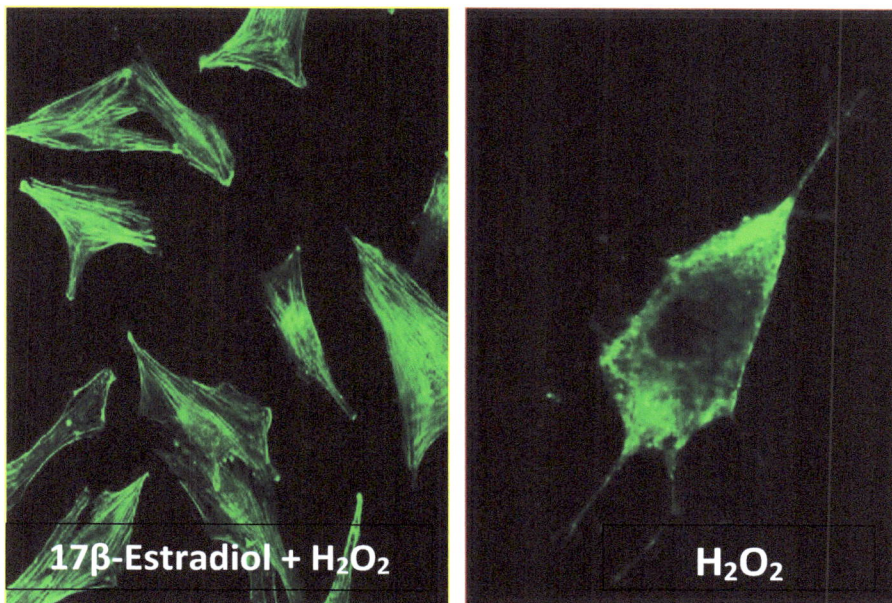

Fig. (1). Observation of actin disorganization in apoptosis by immunocytochemistry assays.
C2C12 cells incubated with estradiol prior to the induction of apoptosis showed an usual actin arrangement. However, when the cells were treated with the apoptosis inducer (H_2O_2) alone, disorganization of the cytoskeleton was observed. Actin was labeled using an anti-actin polyclonal antibody and then using an Alexa Fluor 488-conjugated secondary antibody (green fluorescence).

In agreement, the better muscle composition and performance among the hormone replacement therapy users appeared to be orchestrated by improved regulatory actions on cytoskeleton [26]. For example, estradiol or testosterone by binding their receptors, regulate rapid intracellular cascades such as G proteins, tyrosine kinases, c-Src, small GTPases and kinase pathways, leading to the activation of cytoskeleton remodeling [Review in 9]. The cytoskeleton is the scaffolding and connector of that intricate network. Then, during menopause and andropause, when the sex hormone levels are low, it is conceivable that the cytoskeleton is more prone to be injured by ROS and that all of its cellular interactions are therefore affected.

CONSENT FOR PUBLICATION

Not applicable.

CONFLICT OF INTEREST

The authors confirm that this chapter contents have no conflict of interest.

ACKNOWLEDGEMENTS

National University of the South Argentina and National Research Council of Argentina (CONICET).

REFERENCES

[1]   Huber F, Boire A, López MP, Koenderink GH. Cytoskeletal crosstalk: when three different personalities team up. Curr Opin Cell Biol 2015; 32: 39-47.
        [http://dx.doi.org/10.1016/j.ceb.2014.10.005] [PMID: 25460780]

[2]   Berthier C, Blaineau S. Supramolecular organization of the subsarcolemmal cytoskeleton of adult skeletal muscle fibers. A review. Biol Cell 1997; 89(7): 413-34.
        [http://dx.doi.org/10.1016/S0248-4900(97)89313-6] [PMID: 9561721]

[3]   Pardo JV, Siliciano JD, Craig SW. A vinculin-containing cortical lattice in skeletal muscle: transverse lattice elements ("costameres") mark sites of attachment between myofibrils and sarcolemma. Proc Natl Acad Sci USA 1983; 80(4): 1008-12.
        [http://dx.doi.org/10.1073/pnas.80.4.1008] [PMID: 6405378]

[4]   Ananthakrishnan R, Ehrlicher A. The forces behind cell movement. Int J Biol Sci 2007; 3(5): 303-17.
        [http://dx.doi.org/10.7150/ijbs.3.303] [PMID: 17589565]

[5]   White SR, Williams P, Wojcik KR, *et al.* Initiation of apoptosis by actin cytoskeletal derangement in human airway epithelial cells. Am J Respir Cell Mol Biol 2001; 24(3): 282-94.
        [http://dx.doi.org/10.1165/ajrcmb.24.3.3995] [PMID: 11245627]

[6]   Mills JC, Stone NL, Pittman RN. Extranuclear apoptosis. The role of the cytoplasm in the execution phase. J Cell Biol 1999; 146(4): 703-8.
        [http://dx.doi.org/10.1083/jcb.146.4.703] [PMID: 10459006]

[7]   Povea-Cabello S, Oropesa-Ávila M, de la Cruz-Ojeda P, *et al.* Dynamic Reorganization of the Cytoskeleton during Apoptosis: The Two Coffins Hypothesis. Int J Mol Sci 2017; 18(11): 2393.
        [http://dx.doi.org/10.3390/ijms18112393] [PMID: 29137119]

[8]   Oropesa M, de la Mata M, Maraver JG, *et al.* Apoptotic microtubule network organization and maintenance depend on high cellular ATP levels and energized mitochondria. Apoptosis 2011; 16(4): 404-24.
        [http://dx.doi.org/10.1007/s10495-011-0577-1] [PMID: 21311976]

[9]   Sanchez AM, Flamini MI, Polak K, *et al.* Actin cytoskeleton remodelling by sex steroids in neurones. J Neuroendocrinol 2012; 24(1): 195-201.
        [http://dx.doi.org/10.1111/j.1365-2826.2011.02258.x] [PMID: 22103470]

[10]   Häcker G. The morphology of apoptosis. Cell Tissue Res 2000; 301(1): 5-17.
        [http://dx.doi.org/10.1007/s004410000193] [PMID: 10928277]

[11]   Huxley HE. The mechanism of muscular contraction. Science 1969; 164(3886): 1356-65.
        [http://dx.doi.org/10.1126/science.164.3886.1356] [PMID: 4181952]

[12]   Izdebska M, Zielińska W, Grzanka D, Gagat M. The role of actin dynamics and actin-binding proteins expression in epithelial-to-mesenchymal transition and its association with cancer progression and evaluation of possible therapeutic targets. BioMed Res Int 2018; 2018: 4578373.
        [http://dx.doi.org/10.1155/2018/4578373] [PMID: 29581975]

[13]   Odaka C, Sanders ML, Crews P. Jasplakinolide induces apoptosis in various transformed cell lines by a caspase-3-like protease-dependent pathway. Clin Diagn Lab Immunol 2000; 7(6): 947-52.
        [PMID: 11063504]

[14]   Suria H, Chau LA, Negrou E, Kelvin DJ, Madrenas J. Cytoskeletal disruption induces T cell apoptosis by a caspase-3 mediated mechanism. Life Sci 1999; 65(25): 2697-707.

[http://dx.doi.org/10.1016/S0024-3205(99)00538-X] [PMID: 10622279]

[15]   Pollard TD, Cooper JA. Actin and actin-binding proteins. A critical evaluation of mechanisms and functions. Annu Rev Biochem 1986; 55: 987-1035.
[http://dx.doi.org/10.1146/annurev.bi.55.070186.005011] [PMID: 3527055]

[16]   Desouza M, Gunning PW, Stehn JR. The actin cytoskeleton as a sensor and mediator of apoptosis. Bioarchitecture 2012; 2(3): 75-87.
[http://dx.doi.org/10.4161/bioa.20975] [PMID: 22880146]

[17]   Franklin-Tong VE, Gourlay CW. A role for actin in regulating apoptosis/programmed cell death: evidence spanning yeast, plants and animals. Biochem J 2008; 413(3): 389-404.
[http://dx.doi.org/10.1042/BJ20080320] [PMID: 18613816]

[18]   Zheng S, Huang J, Zhou K, *et al.* 17β-Estradiol enhances breast cancer cell motility and invasion *via* extra-nuclear activation of actin-binding protein ezrin. PLoS One 2011; 6(7): e22439.
[http://dx.doi.org/10.1371/journal.pone.0022439] [PMID: 21818323]

[19]   Fujita H, Allen PG, Janmey PA, *et al.* Induction of apoptosis by gelsolin truncates. Ann N Y Acad Sci 1999; 886: 217-20.
[http://dx.doi.org/10.1111/j.1749-6632.1999.tb09420.x] [PMID: 10667223]

[20]   Koya RC, Fujita H, Shimizu S, *et al.* Gelsolin inhibits apoptosis by blocking mitochondrial membrane potential loss and cytochrome c release. J Biol Chem 2000; 275(20): 15343-9.
[http://dx.doi.org/10.1074/jbc.275.20.15343] [PMID: 10809769]

[21]   Ting HJ, Chang C. Actin associated proteins function as androgen receptor coregulators: an implication of androgen receptor's roles in skeletal muscle. J Steroid Biochem Mol Biol 2008; 111(3-5): 157-63.
[http://dx.doi.org/10.1016/j.jsbmb.2008.06.001] [PMID: 18590822]

[22]   Nishimura K, Ting HJ, Harada Y, *et al.* Modulation of androgen receptor transactivation by gelsolin: a newly identified androgen receptor coregulator. Cancer Res 2003; 63(16): 4888-94.
[PMID: 12941811]

[23]   Lim JJ, Ngah WZ, Mouly V, Abdul Karim N. Reversal of myoblast aging by tocotrienol rich fraction posttreatment. Oxid Med Cell Longev 2013; 2013: 978101.
[http://dx.doi.org/10.1155/2013/978101] [PMID: 24349615]

[24]   Vasconsuelo A, Milanesi L, Boland R. 17β-estradiol abrogates apoptosis in murine skeletal muscle cells through estrogen receptors: role of the phosphatidylinositol 3-kinase/Akt pathway. J Endocrinol 2008; 196(2): 385-97.
[http://dx.doi.org/10.1677/JOE-07-0250] [PMID: 18252962]

[25]   Pronsato L, Ronda AC, *et al.* Protective role of 17b-estradiol and testosterone in apoptosis of skeletal muscle. Actual Osteol 2010; 2: 45-8.

[26]   Ronkainen PH, Pöllänen E, Alén M, *et al.* Global gene expression profiles in skeletal muscle of monozygotic female twins discordant for hormone replacement therapy. Aging Cell 2010; 9(6): 1098-110.
[http://dx.doi.org/10.1111/j.1474-9726.2010.00636.x] [PMID: 20883525]

CHAPTER 7

Apoptosis As Cause Of Sarcopenia: Hormonal Regulation

Andrea Vasconsuelo*

Instituto de Ciencias Biológicas y Biomédicas del Sur (INBIOSUR), Universidad Nacional del Sur- CONICET, Bahía Blanca, Argentina

Abstract: The muscle mass declines with age, affecting the independence in the elderly. This process known as sarcopenia and the acceleration of myocytes loss through apoptosis might signify the main causative.

Skeletal muscle tissue increases its size and shows a notable capacity to adapt to injury, due the existence of an undifferentiated group of myogenic-specific precursor cells, called satellite cells. The knowledge of mechanism that driven the apoptosis in satellite cells represents an important base to understand the etiology of sarcopenia.

This chapter centers on the potential impact that the estrogen- and testosterone-regulation of satellite cell function has in elderly skeletal muscle, highlighting in the role that both steroids have on apoptotic signaling in myoblasts.

Keywords: Apoptosis Pathways, Bcl-2 Family, Caspases, Muscle Structure, Skeletal Muscle.

INTRODUCTION

Aging, an predictable biological process, linked with a loss of tissue and organ function in a gradual fashion [1]. Molecularly, the aging process implies a progressive worsening of biomolecules to yield a variety of pathological outcomes, such as cancer, neurodegenerative diseases, sarcopenia, and hepatic dysfunctions [2 - 4]. Regards skeletal muscle, as the human body ages, the muscle mass declines and this process is accompanied by a reduction of strength and/or physical performance affecting the daily activities and freedom in the elderly [5, 6]. The process that begins near of age 30 with an important acceleration post age 60 is often referred to as "sarcopenia", from Greek sarx: flesh and penia: deficiency [7]. However, there is still no widely established definition of this condition. The more collective clinical definition was developed in 2009–2010 by

* **Corresponding author Andrea A. Vasconsuelo:** Instituto de Ciencias Biológicas y Biomédicas del Sur (INBIOSUR), Universidad Nacional del Sur- CONICET, Bahía Blanca, Argentina; E-mail: avascon@criba.edu.ar

the European Working Group on Sarcopenia in Older People (EWGSOP) that defined sarcopenia as "a geriatric syndrome characterized by involuntary progressive and generalized loss of skeletal muscle mass and strength, resulting in physical disability, poor quality of life and death [6]. Actually, is recognized as a key factor responsible for the occurrence of frailty and negative health outcomes [8]. The detailed pathogenesis of sarcopenia have not yet been fully determined. Similar to other age related conditions, sarcopenia is characterized by a multifactorial etiology, for example it has been associated to physical inactivity [9], to motor neuronal loss due to aging, to 25(OH) vitamin D low levels found in olders = older persons or the elderly to high levels of catabolic cytokines [10], to nutritional disorders [11] or to defects in aged mitochondria, among others [Review in 12].

However, the specific input of each of these factors and the molecular mechanisms triggered by those conditions that finally lead to fiber loss, are still unrevealed. Regards the role of mitochondria in sarcopenia, it is known that increased mitochondrial fusion during aging results in giant mitochondria that are bioenergetically inefficient. Moreover, the mitochondrial morphology changes with age, an increased volume and fewer cristae have been observed in this organelle in older cells [13, 14]. Then, this giant organelle generate more reactive oxygen species (ROS) and the primary target of ROS injury is the mitochondria itself showing a greater propensity to induce muscle cell death by apoptosis [15 - 17]. Summarizing, mitochondrial morphology and functions are attacked by chronic exposure to mitochondrial ROS during aging. Indeed, the free radical principle of Harman (1956) [18] and later refined by Miquel *et al.* [17] suggests that mitochondrial ROS are the key contributors to the damage of biological macromolecules. Causing a permanent cell impairment [19 and references therein,18], *via* distressing the mitochondrial genome, the cellular energy production, the induction/regulation of apoptosis, and other mitochondrial functions. Terman *et al.,* (2003) [20] suggest that oxidative stress and subsequent mitochondrial swelling appears to be responsible for the enlargement of the mitochondria. Although the mechanisms responsible for all age-dependent alterations of the organelle are not fully understood, it is known that these alterations induce apoptosis of muscle cells.

Age appears to affect both mitochondrial DNA (mtDNA) content and integrity. Numerous studies, report that mtDNA copy number decreases in aged human skeletal muscle cells due, at least in part, to oxidative damage [Review in 21]. The age-associated decline in mtDNA copy number tends to be greater in more oxidative fibers [22]. However also exist data indicating that, the ageing don't affect the mtDNA copy number [23]. In chapter 1, we describe that myofibers can be grouped into a slow-contracting/fatigue-resistant type and a fast-contracting/ fatigue-susceptible type, according their physiological properties [24]. The slow-contracting/fatigue resistant type, have higher mitochondrial content. Then this class of myofiber could be more susceptible to apoptosis triggering a massive mitochondrial apoptotic signal by bioenergetically inefficient mitochondria due aging. All these data suggest that mitochondria could play a key role in sarcopenia development.

Although the causes of sarcopenia above enumerated are non-endocrine, hormones regulate several of them directly or indirectly. Then, the doubt arises that there is a degree of endocrine control over all possible causes of sarcopenia. Actually, there is a large body of evidence indicating that sarcopenia is closely related to the decline in sex hormones that occurs with aging [12, 25]. In concurrence, the decrease in muscle mass presents differences according to gender [26].

The changes in hormonal status mainly the decline in estrogen levels in menopause contributes to the decrease in bone mass density, the redistribution of subcutaneous fat to the visceral area, the increased risk of cardiovascular disease and have a direct negative effect on muscle mass [27]. In menopause an accelerated decline in muscle mass and changes in quality of muscle tissue, reducing quality of life, have been reported [28]. Certainly, postmenopausal women had twice the amount of intramuscular fat, compared to younger women [29]. Similar to women, in men free testosterone levels decrease in elderly (andropause) and this decline matches with the decrease in muscle mass and strength [30]. Regards supplementation therapy it has been shown that the results in men are variable as some studies have observed an increase in muscle mass while others have not [31], however these studies do not consider whether there are modifications in the expression of androgen receptor (AR) as possible causal of the different responses. On the other hand, respect women; there is a growing body of evidence describing the positive effects and mechanisms behind estrogen and hormone therapy (HTR) effects on muscle system [Review in 32]. Nevertheless, the molecular mechanisms by which a decrease in steroid sexual hormone levels may have a negative effect on muscle mass are not well understood. In the present chapter, we discuss the current knowledge regarding the molecular mechanism triggered by androgen and estrogen at the mitochondria and their effects in elderly in relation to sarcopenia associated with the deregulation of sexual hormones levels.

Mitochondria and Sexual Hormones

For a long time, mitochondria have been known as the 'powerhouse' of the cell, producing the energy required for cell metabolism by oxidative phosphorylation [33]. It is now accepted that mitochondria are also involved in numerous other physiological processes such as apoptosis, in which play a key role [34]. Furthermore, this polyfunctional organelle also contributes in cellular signaling [35]. Someway, these mitochondrial processes are regulated by steroid and thyroid hormones in the progression of their actions on metabolism, growth, and development [36]. Regards sex steroid hormones have been shown that, while mitochondria mediate sex hormone production, sex steroid hormones can also regulate mitochondrial functions indirectly through intracellular messengers or directly through specific receptors located in the organelle. In previous chapters we described that estrogen and testosterone classically task by binding to their receptors: estrogen receptors (ER) and androgen receptors (AR), respectively. Also, we mentioned that has been well documented that both ER and AR are located in mitochondria of mammalian cells, including skeletal muscle cell [37 - 43]. In keeping with this subcellular localization, studies have been performed to evaluate the actions of 17β-estradiol (E2) and testosterone (T) through their receptors on the organelle. The results obtained further confirm that mitochondria are an estrogen and androgen target. Thus, there is evidence that steroid hormones affects the electron transport chain, gene expression and mitochondrial morphology, among others [38, 44 - 46]. Moreover, linked with apoptosis has been shown that these steroids modulate the mitochondrial membrane potential and in consequence the opening of the mitochondrial permeability transition pore (MPTP) with release of mitochondrial components [40, 44 - 47].

As example of indirect action of estrogen on mitochondria, we could mention the difference in longevity between males and females in many species. This has been attributed to a lower ROS production by mitochondria from females when compared with those of males and this is because estradiol up-regulates the nuclear expression of antioxidant genes whose products are directed towards mitochondria. Antioxidants enzymes such as manganese superoxide dismutase (MnSOD) and glutathione peroxidase (GSH-Px) are examples of this [48]. In agreement, we observed that estradiol protects the skeletal muscle cell against an apoptotic stimulus, regulating mitochondrial enzymes; we found that H_2O_2 (apoptosis inducer) treatment reduced the levels of MnSOD suggesting that this could be a factor contributing to muscle apoptosis. Preincubation with E2 diminished the negative effects on MnSOD; the hormone increased the enzyme expression and activity. Moreover, the ER antagonist fulvestrant blocked the estrogen effects on MnSOD in skeletal muscle cell, involving the ER in this antiapoptotic mechanism [45]. In similar fashion, in various cellular models, low levels of testosterone were found to be associated with oxidative stress induced by decreased expression of antioxidant enzymes such as MnSOD, GSH-Px, and catalase [49]. Actually, has been evidenced a direct interaction of sex steroids with mitochondria, showing an antioxidant and protective effect of estradiol on isolated mitochondria resulting in inhibition of cytochrome c release. Thus, the hormone prevents the trigger of the mitochondrial pathway of apoptosis [50].

Despite the clear action of sexual hormones on mitochondria, other mechanisms that related these hormones with sarcopenia has suggested. For example, the decrease in estrogen levels during menopause may be associated with an increase in pro-inflammatory cytokines, such as tumor necrosis factor alpha (TNF-α) or interleukine-6 (IL-6), which might be implicated in the apparition of sarcopenia, since both cytokines contributes to muscle catabolism [51].

Likewise, dehydroepiandrosterone (DHEA), an intermediate produced during the biosynthesis of steroids hormones that can transform into sex steroid such as androgens and estrogens, has been associated with muscle mass loss. Among the roles of DHEA in the human body, it may contribute to the increase in muscle mass [37]. Circulating levels of DHEA decline with age, especially at menopause [2]. This decline in DHEA has been shown to be associated with a decrease in muscle mass and physical performance [37]. However, Abbasi *et al.* [38] did not observe a relationship between DHEA levels and body composition in women aged 60 years and older. Furthermore, in elderly individuals, DHEA replacement showed no improvement in physical performance and body composition [39].

Apoptosis As Cause Of Sarcopenia

With the aim of develop strategies to prevent and treat sarcopenia, the risk factors, the molecular mechanism involved and the etiology of this pathology must be identified. Although sarcopenia is the result of both intrinsic factors including changes at the molecular and cellular levels, and extrinsic or environmental/ lifestyle behaviors such as physical inactivity, smoking and poor diet [52], here it will focus in molecular and cellular mechanisms. As was described, epidemiological studies have suggested multiple contributing factors such as gradual loss of motor neurons, reduced growth factors signaling and protein synthesis, elevated amounts of circulating cytokines and increased oxidative stress among others [Review in 53]. Interesting, muscle proteolysis and apoptosis are closely interconnected given that apoptotic signaling is prerequisite for and precedes protein degradation during muscle wasting [54]. Other proteolytic systems have been involved in muscle degradation, such as autophagy, calcium-activated proteases, and the ubiquitin-proteasome system [55, 56]. However, evidence indicates that an age-related acceleration of myocytes loss *via* apoptosis might represent a main mechanism driving the onset and progression of sarcopenia [57, 58]. For example, has been shown that downregulation of the apoptotic pathway can decrease the deterioration in muscle mass and function in aged animals [59, 60]. Furthermore, upregulation of the apoptotic pathway has been identified in premature aging models including mice lacking the antioxidant enzyme superoxide dismutase 1 (SOD1) that exhibit accelerated sarcopenia [61]. Likewise, numerous studies have exposed that the extent of apoptotic DNA fragmentation increases in skeletal muscle in elderly, thus paralleling the progress of sarcopenia. Moreover, this fragmentation is likely responsible of the decrease in transcriptional efficiency present in sarcopenia [Review in 62]. Respect caspases, their activation vary depending on the experimental models used [63]. In agreement, Baker and Hepple using plantaris muscles from rats shown that the apoptosis inducer factor (AIF) and Caspases 3, 8, and 9 gene expression increased with aging in proportion to the degree of sarcopenia, whereas BAX, Bcl-2, and Apaf-1 decline with aging [64]. As observed in other cellular systems, apoptosis may be initiated in muscle cells by external stimuli trough cell membrane death receptors as well as by internal stimuli through the mitochondrial apoptotic pathway [Review in 65]. Previously we describe that skeletal muscle represents a unique tissue with respect to apoptotic process at mitochondrial and nuclear level. Muscle fibers are multinuclear cells and the death of one nucleus by apoptosis, although generally has a negative effect on gene expression in the surrounding cytoplasm, does not lead to the death of the whole muscle fiber. This makes us suppose the nuclei present in the muscle fibers as autonomous entities, capable of recognizing specific signaling since the stimulus that triggered the apoptosis in a certain nucleus is not capable of triggering it in near nuclei that share the same cytoplasm. Reinforcing the concept of "nuclear autonomy", has been demonstrated that within an individual muscle fiber, not all nuclei are transcriptionally equivalent, but that gene expression seems to be regulated independently between nuclei [Review in 58].

Regards mitochondria, skeletal muscle present different types of this organelle, subject to the type of muscle fiber and activity [66], or subject to cellular localization (located directly underneath the sarcolemma or intermyofibrillar). Then, arise the possibility of differential response of the organelle to diverse stimuli. Depending on their localization, has been shown that mitochondria possess different functional, compositional and biochemical properties [review in 65]. Probably, those different compositional implies different expression of the androgen and estrogen receptors and therefore the susceptibility to hormonal regulation of apoptosis is variable depending on the mitochondrial localization.

We have already mentioned that sarcopenia is a multifactorial disease, and it is difficult to establish the degree of contribution of each factor/mechanism to the development of the pathology, since many of these occur simultaneously; as well as they are cause or consequence among themselves, of the participation of the other. For example, the effects of age-related oxidative stress in skeletal muscle can determine not only the imbalance between protein synthesis and degradation, but also mitochondrial dysfunction and then induce apoptosis. In addition, low levels of sex steroids due to aging do not protect against age-related oxidative stress that make the situation worse. This complex system involved in the homeostasis of skeletal muscle prompt research into the role of apoptosis and their regulation factors, in sarcopenia.

Apoptosis Of Satellite Cells

Adult skeletal muscle increases its size and shows a remarkable capacity to adapt to trauma and injury. This regenerative capacity of skeletal muscle is reliant on a small undifferentiated group of myogenic-specific precursor cells, called satellite cells (SC) [67, 68]. Indeed, myonuclei in skeletal muscle are postmitotic and cannot replicate. Therefore, any increase in myonuclear number required for growth or repair of damaged muscle depends on SC, which were first described by Mauro (1961) with an electron microscope [69]. They owe their name to their localization under the basement membrane but outside the plasma membrane of the muscle fiber. In addition, their colocalization with blood vessels [70, 71] places SC in an optimal location to respond to intrinsic signals from both the skeletal muscle fiber itself and from changes in the systemic environment.

SC are activated in response to both physiological and pathological stimuli, such as exercise, injury and degenerative diseases or during development. Once quiescent SC are activated, up-regulate the expression of c-Met, Pax-7, M-cadherin and activate the expression of factors involved in the specification of the myogenic program such as myf-5 and MyoD. Then, start to proliferate and they are often referred to as myogenic precursor cells or myoblasts [72], the process of formation of new myonuclei, is known as myogenesis [Review in 73].

As was describe before, the number of myofibers declines with aging and due that SC are the only cells producing myofibres; thus, their relevance in sarcopenia is implicated. SC number is not constant throughout life and depends on age and muscle fiber type. These cells are most plentiful during development contributing to muscle growth, and decay in number thereafter. For instance, in rats it has been shown that the number of SC was reduced by 22% in older animals compared with their younger counterparts [74]. In humans, rat and mice has been observed that the number of SC also declines in the ageing [75, 76]. However, other authors propose that it is controversial whether satellite cells decrease in number in aging skeletal muscle [77]. On the other hand, most studies indicate that during aging satellite cells display reduced proliferative response after damage and reduced regenerative capacity. In aging muscle, satellite cells may also display a tendency to adopt alternate lineages, showing fibrogenic potential that could contribute to muscle fibrosis [78].

Although the etiology of sarcopenia is complex, the correlation during aging between the loss of SC activity, and impaired muscle regenerative capacity has led to the hypothesis that the death of SC or the loss of its regenerative capacity is also a cause of sarcopenia [46]. Additionally, if we consider the cellular microenvironment and particularly in the case of the SC that, as previously mentioned, are in an optimal location to sense changes in their environment, perhaps the aged muscle is a prohibitive atmosphere for SC activation and accurate function. Nevertheless, Fry *et al.* showed that SC are dispensable for the onset and progress of sarcopenia; they show that skeletal muscle apparently employs cellular mechanisms that do not necessarily require stem cell participation for tissue maintenance [79]. This discrepancy appeared to depend on the types of tissues analyzed and the animal model used [Review in 80]. In addition, SC are a heterogeneous population in terms of cell size and clonogenic potential [81]. Several studies have raised the possibility that SC are a heterogeneous mixture of stem cells and committed myogenic progenitors [82, 83]. Relevant investigations demonstrated that an important percentage (near of 80%) of SC divide quickly and yield myonuclei to growing fibers in rat skeletal muscle. The lasting 20% of the cells enter into the G0 phase regarding as backup cells, thus dividing more slowly [83]. Interesting, the 80% of SC express the CD34, Myf5 and M-cadherin markers while the rest of cells do not express them [84]. Really, SC express specific markers throughout the different phases of muscle regeneration [review in 85]. Although further studies will be necessary to achieve conclusive results on the roles of SC in sarcopenia, the most of the available evidence indicates that these cells play a central role in the development of this pathology.

Here, we focuses on the potential impact that the estrogen and testosterone has on SC functions, in aged skeletal muscle. The implications of the sex steroids effects on SC are particularly important for matured persons, where circulating estrogen and testosterone levels decay and in consequence, their favorable influences on skeletal muscle tissue may diminish.

Of significance, it has been shown that there are gender variances in the reactions of SC to the injury suggesting that SC activation could be regulated by E2 and/or T. In fact, these steroids affect several muscle genes such as Pax-7, MyoD and myogenin [46]. Compatible with these observations, it is known that satellite cells express the androgen receptor (AR) [86, 87] as well as the female hormone receptors (ERs); ratifying that SC are targets for androgen and estrogen actions [Review in 46].

The murine myoblasts cell line, named C2C12, are from satellite cells. The C2C12 cell line behavior corresponds to that of progenitor lineage. This cell line is a subclone of C2 myoblasts [88] which naturally proliferate, differentiate and synthesize typical muscle proteins [89, 90]. Thus, C2C12 skeletal myoblast cells are comparable to satellite cells in muscle fibers [91]. As was described in previous chapters in the C2C12 cells, biochemical, immunological and molecular data support the mitochondrial-microsomal localization of ERα [92]. This data were confirmed using specific ERα-siRNA and by RT-PCR assays [92]. Likewise, in the same cell line, ERβ was detected mainly in mitochondria and in lower amounts in the cytosolic fraction, these results were established using conventional and confocal microscopy, and corroborated after transient transfection with specific ERβ-siRNAs and also by RT-PCR [43]. These results show that ERβ localizes differently to ERα in C2C12 cells [43; review in 93]. Also, biochemical and immunological data demonstrated mitochondrial and microsomal localization of AR in skeletal muscle cells [38].

Previously in this section we described the relation of sarcopenia and apoptosis. A big body of evidence, indicate that androgens and estrogens regulate apoptosis *via* different intracellular signaling pathways, depending on factors such as cell type, apoptosis stimuli, hormone concentration and cellular environment [review in 93]. In the skeletal muscle myoblasts, T and E2 protect against apoptosis trigger by hydrogen peroxide [44, 94]. For example, it has been demonstrated that classic events of apoptosis such as DNA cleavage, cytoskeleton disorganization, mitochondrial reorganization/dysfunction, loss of mitochondrial membrane potential with cytochrome c release induced by H_2O_2, are abolished when myoblasts are previously exposed to androgen or estrogen [review in 93]. Thus, this response of myoblasts suggest that the decline in E2 and/or T make these cells more susceptible to apoptotic cell death.

Part of molecular mechanism of antiapoptotic effects of T or E2 on myoblasts have also been identified. At short times of exposure to hydrogen peroxide, cells display a defense response concerning ERK2, Akt and Bad phosphorylation and a growth of HSP70 levels. However, at extended treatment times with H_2O_2, dephosphorylation of these proteins, loss of the mitochondrial membrane potential, cytochrome c release, PARP cleavage and destruction of genetic material occur. When cells are treated with T prior to apoptosis induction, Bad inactivation, growth in actin levels, traffic of HSP90 to mitochondria, inhibition of the beating of the mitochondrial membrane potential, and a decrease in PARP cleavage and Bax levels, are detected [47, 95]. This antiapoptotic action of T is mediated by AR [38], indicating a specific survival action of the steroid involving the activation of different cellular signaling pathways. These results expose that at least the intrinsic pathway, is under control of male hormone in skeletal muscle myoblast. Similarly, E2 prevents apoptosis in murine myoblasts through ERs concerning MAPKs and PI3K/Akt pathways, which phosphorylates and inhibits proapoptotic members of the Bcl-2 family. Analogous to T, in the defensive effects of E2 on mitochondria is implicated the PI3K/Akt pathway. But dissimilar T, E2 antiapoptotic action comprises HSP27 but not HSP90 [review in 93], indicating specificity of signalling targets for each hormone. Possibly these differences in the chaperones involved are driven by characteristic domains in ERs and ARs. For that reason, these HSPs are capable to act together with ERs or AR stabilizing them [96, 97].

Among the protective actions of both steroids at mitochondrial level against the apoptotic inducer, it has been demonstrated an increased mitochondrial manganese superoxide dismutase protein expression and activity, opening of mitochondrial permeability transitory pore (MPTP) inhibition and consequently protection of mitochondrial membrane potential [38, 45]. In keeping with these events, both hormones inhibit the translocation of Bax to mitochondria, induced by hydrogen peroxide [45, 95]. The defensive effects of male hormone apparently are specific of skeletal muscle cells since in cardiomyocytes, testosterone therapy has no protective effect in acute muscle injury associated with increased muscle cell death after cardiotoxin treatment [98]. These information are important since, as it was mentioned, growing evidence designates that an age-related deregulation of apoptosis may contribute to the onset and evolution of sarcopenia due to an increase in the proportion of cellular loss. Certainly, an accelerated death of satellite cells in aged muscle could conduce to an impaired muscle regenerative process in the elderly. In addition, the data suggest that altered sex hormone levels could affect the normal response of the cells that maintain and repair skeletal muscle during sarcopenia or other deleterious effects of aging. The fact that the results described above were achieved in satellite cells treated with hydrogen peroxide, strengthen this perception. Indeed, aging can be induced precipitately by oxidative stress, through the accumulation of ROS [99]. Oxidants, such as hydrogen peroxide, may attack numerous kinds of tissue, including skeletal muscle. This tissue is prone to oxidative stress-induced aging as the myofibers are highly oxygen-consuming structures, and the level of ROS formed in skeletal muscle is higher than in other tissues [99, 100]. Undoubtedly, aging is related with extreme ROS levels, which increase mitochondrial injury [review in 101]. Mitochondria might harvest higher oxidants levels in the matured skeletal muscle, disabling satellite cells and then, contributing to the impairment of its restorative role. Thus, C2C12 cells exposed to H_2O_2 resemble an aging-like phenotype similar to aged skeletal muscle [102] and represent a valuable tool to understand the molecular mechanisms concerned in sarcopenia.

This information will expected help to the project of more current therapeutic approaches to preserve muscle mass in the elderly.

CONSENT FOR PUBLICATION

Not applicable.

CONFLICT OF INTEREST

The authors confirm that this chapter contents have no conflict of interest.

ACKNOWLEDGEMENTS

National University of the South Argentina and National Research Council of Argentina (CONICET).

REFERENCES

[1]   Flatt T. A new definition of aging? Front Genet 2012; 3: 148.
        [http://dx.doi.org/10.3389/fgene.2012.00148] [PMID: 22936945]

[2]   Chung JH, Seo AY, Chung SW, *et al.* Molecular mechanism of PPAR in the regulation of age-related inflammation. Ageing Res Rev 2008; 7(2): 126-36.
        [http://dx.doi.org/10.1016/j.arr.2008.01.001] [PMID: 18313368]

[3]   Chung HY, Cesari M, Anton S, *et al.* Molecular inflammation: underpinnings of aging and age-related diseases. Ageing Res Rev 2009; 8(1): 18-30.
        [http://dx.doi.org/10.1016/j.arr.2008.07.002] [PMID: 18692159]

[4]   Seo AY, Hofer T, Sung B, Judge S, Chung HY, Leeuwenburgh C. Hepatic oxidative stress during aging: effects of 8% long-term calorie restriction and lifelong exercise. Antioxid Redox Signal 2006; 8(3-4): 529-38.
        [http://dx.doi.org/10.1089/ars.2006.8.529] [PMID: 16677097]

[5]   Sinha-Hikim I, Roth SM, Lee MI, Bhasin S. Testosterone-induced muscle hypertrophy is associated with an increase in satellite cell number in healthy, young men. Am J Physiol Endocrinol Metab 2003; 285(1): E197-205.
        [http://dx.doi.org/10.1152/ajpendo.00370.2002] [PMID: 12670837]

[6]   Cruz-Jentoft AJ, Baeyens JP, Bauer JM, *et al.* Sarcopenia: European consensus on definition and diagnosis: report of the European Working Group on Sarcopenia in Older People. Age Ageing 2010; 39(4): 412-23.
        [http://dx.doi.org/10.1093/ageing/afq034] [PMID: 20392703]

[7]   Rosenberg IR. Summary comments. Am J Clin Nutr 1989; 50: 1231-3.
        [http://dx.doi.org/10.1093/ajcn/50.5.1231]

[8]   Landi F, Calvani R, Cesari M, *et al.* Sarcopenia: an overview on current definitions, diagnosis and treatment. Curr Protein Pept Sci 2018; 19(7): 633-8.
        [http://dx.doi.org/10.2174/1389203718666170607113459] [PMID: 28595526]

[9]   Szulc P, Duboeuf F, Marchand F, Delmas PD. Hormonal and lifestyle determinants of appendicular skeletal muscle mass in men: the MINOS study. Am J Clin Nutr 2004; 80(2): 496-503.
        [http://dx.doi.org/10.1093/ajcn/80.2.496] [PMID: 15277176]

[10]   Visser M, Pahor M, Taaffe DR, *et al.* Relationship of interleukin-6 and tumor necrosis factor-alpha with muscle mass and muscle strength in elderly men and women: the Health ABC Study. J Gerontol A Biol Sci Med Sci 2002; 57(5): M326-32.
        [http://dx.doi.org/10.1093/gerona/57.5.M326] [PMID: 11983728]

[11]   Dreyer HC, Volpi E. Role of protein and amino acids in the pathophysiology and treatment of

sarcopenia. J Am Coll Nutr 2005; 24(2): 140S-5S.
[http://dx.doi.org/10.1080/07315724.2005.10719455] [PMID: 15798081]

[12]   McKee A, Morley JE, Matsumoto AM, Vinik A. Sarcopenia: An Endocrine Disorder? Endocr Pract 2017; 23(9): 1140-9.
[http://dx.doi.org/10.4158/EP171795.RA] [PMID: 28704095]

[13]   Wilson PD, Franks LM. The effect of age on mitochondrial ultrastructure. Gerontologia 1975; 21(2): 81-94.
[http://dx.doi.org/10.1159/000212035] [PMID: 1158107]

[14]   Ozawa T. Genetic and functional changes in mitochondria associated with aging. Physiol Rev 1997; 77(2): 425-64.
[http://dx.doi.org/10.1152/physrev.1997.77.2.425] [PMID: 9114820]

[15]   Calvani R, Joseph AM, Adhihetty PJ, *et al.* Mitochondrial pathways in sarcopenia of aging and disuse muscle atrophy. Biol Chem 2013; 394(3): 393-414.
[http://dx.doi.org/10.1515/hsz-2012-0247] [PMID: 23154422]

[16]   Sastre J, Pallardó FV, Plá R, *et al.* Aging of the liver: age-associated mitochondrial damage in intact hepatocytes. Hepatology 1996; 24(5): 1199-205.
[http://dx.doi.org/10.1002/hep.510240536] [PMID: 8903398]

[17]   Miquel J, Economos AC, Fleming J, Johnson JE Jr. Mitochondrial role in cell aging. Exp Gerontol 1980; 15(6): 575-91.
[http://dx.doi.org/10.1016/0531-5565(80)90010-8] [PMID: 7009178]

[18]   Harman D. Aging: a theory based on free radical and radiation chemistry. J Gerontol 1956; 11(3): 298-300.
[http://dx.doi.org/10.1093/geronj/11.3.298] [PMID: 13332224]

[19]   Kowaltowski AJ, de Souza-Pinto NC, Castilho RF, Vercesi AE. Mitochondria and reactive oxygen species. Free Radic Biol Med 2009; 47(4): 333-43.
[http://dx.doi.org/10.1016/j.freeradbiomed.2009.05.004] [PMID: 19427899]

[20]   Terman A, Dalen H, Eaton JW, Neuzil J, Brunk UT. Mitochondrial recycling and aging of cardiac myocytes: the role of autophagocytosis. Exp Gerontol 2003; 38(8): 863-76.
[http://dx.doi.org/10.1016/S0531-5565(03)00114-1] [PMID: 12915208]

[21]   Courtney M. Johannsen, and Eric Ravussin, "Skeletal muscle mitochondria and aging: A review," J Aging Res 2012; vol. 2012(Article ID 194821): 20 pages.

[22]   Barazzoni R, Short KR, Nair KS. Effects of aging on mitochondrial DNA copy number and cytochrome c oxidase gene expression in rat skeletal muscle, liver, and heart. J Biol Chem 2000; 275(5): 3343-7.
[http://dx.doi.org/10.1074/jbc.275.5.3343] [PMID: 10652323]

[23]   Barrientos A, Casademont J, Cardellach F, *et al.* Qualitative and quantitative changes in skeletal muscle mtDNA and expression of mitochondrial-encoded genes in the human aging process. Biochem Mol Med 1997; 62(2): 165-71.
[http://dx.doi.org/10.1006/bmme.1997.2647] [PMID: 9441868]

[24]   Baldwin KM, Haddad F. Effects of different activity and inactivity paradigms on myosin heavy chain gene expression in striated muscle. J Appl Physiol 2001; 90(1): 345-57.
[http://dx.doi.org/10.1152/jappl.2001.90.1.345] [PMID: 11133928]

[25]   Morley JE, Malmstrom TK. Frailty, sarcopenia, and hormones. Endocrinol Metab Clin North Am 2013; 42(2): 391-405.
[http://dx.doi.org/10.1016/j.ecl.2013.02.006] [PMID: 23702408]

[26]   Janssen I, Heymsfield SB, Wang ZM, Ross R. Skeletal muscle mass and distribution in 468 men and women aged 18-88 yr. J Appl Physiol 2000; 89(1): 81-8.
[http://dx.doi.org/10.1152/jappl.2000.89.1.81] [PMID: 10904038]

[27]   Carr MC. The emergence of the metabolic syndrome with menopause. J Clin Endocrinol Metab 2003; 88(6): 2404-11.
[http://dx.doi.org/10.1210/jc.2003-030242] [PMID: 12788835]

[28]   Aloia JF, McGowan DM, Vaswani AN, Ross P, Cohn SH. Relationship of menopause to skeletal and muscle mass. Am J Clin Nutr 1991; 53(6): 1378-83.
[http://dx.doi.org/10.1093/ajcn/53.6.1378] [PMID: 2035465]

[29]   Jubrias SA, Odderson IR, Esselman PC, Conley KE. Decline in isokinetic force with age: muscle cross-sectional area and specific force. Pflugers Arch 1997; 434(3): 246-53.
[http://dx.doi.org/10.1007/s004240050392] [PMID: 9178622]

[30]   van den Beld AW, de Jong FH, Grobbee DE, Pols HA, Lamberts SW. Measures of bioavailable serum testosterone and estradiol and their relationships with muscle strength, bone density, and body composition in elderly men. J Clin Endocrinol Metab 2000; 85(9): 3276-82.
[PMID: 10999822]

[31]   Gruenewald DA, Matsumoto AM. Testosterone supplementation therapy for older men: potential benefits and risks. J Am Geriatr Soc 2003; 51(1): 101-15.
[http://dx.doi.org/10.1034/j.1601-5215.2002.51018.x] [PMID: 12534854]

[32]   Tiidus PM. Benefits of estrogen replacement for skeletal muscle mass and function in post-menopausal females: evidence from human and animal studies. Eurasian J Med 2011; 43(2): 109-14.
[http://dx.doi.org/10.5152/eajm.2011.24] [PMID: 25610174]

[33]   McBride HM, Neuspiel M, Wasiak S. Mitochondria: more than just a powerhouse. Curr Biol 2006; 16(14): R551-60.
[http://dx.doi.org/10.1016/j.cub.2006.06.054] [PMID: 16860735]

[34]   Nikoletopoulou V, Markaki M, Palikaras K, Tavernarakis N. Crosstalk between apoptosis, necrosis and autophagy. Biochim Biophys Acta 2013; 1833(12): 3448-59.
[http://dx.doi.org/10.1016/j.bbamcr.2013.06.001] [PMID: 23770045]

[35]   Brookes PS, Pinner A, Ramachandran A, *et al.* High throughput two-dimensional blue-native electrophoresis: a tool for functional proteomics of mitochondria and signaling complexes. Proteomics 2002; 2(8): 969-77.
[http://dx.doi.org/10.1002/1615-9861(200208)2:8<969::AID-PROT969>3.0.CO;2-3]

[36]   Green DR, Reed JC. Mitochondria and apoptosis. Science 1998; 281(5381): 1309-12.
[http://dx.doi.org/10.1126/science.281.5381.1309] [PMID: 9721092]

[37]   Solakidi S, Psarra AM, Nikolaropoulos S, Sekeris CE. Estrogen receptors alpha and beta (ERalpha and ERbeta) and androgen receptor (AR) in human sperm: localization of ERbeta and AR in mitochondria of the midpiece. Hum Reprod 2005; 20(12): 3481-7.
[http://dx.doi.org/10.1093/humrep/dei267] [PMID: 16123086]

[38]   Pronsato L, Boland R, Milanesi L. Non-classical localization of androgen receptor in the C2C12 skeletal muscle cell line. Arch Biochem Biophys 2013; 530(1): 13-22.
[http://dx.doi.org/10.1016/j.abb.2012.12.011] [PMID: 23262317]

[39]   Monje P, Boland R. Subcellular distribution of native estrogen receptor alpha and beta isoforms in rabbit uterus and ovary. J Cell Biochem 2001; 82(3): 467-79.
[http://dx.doi.org/10.1002/jcb.1182] [PMID: 11500923]

[40]   Chen JQ, Yager JD. Estrogen's effects on mitochondrial gene expression: mechanisms and potential contributions to estrogen carcinogenesis. Ann N Y Acad Sci 2004; 1028: 258-72.
[http://dx.doi.org/10.1196/annals.1322.030] [PMID: 15650251]

[41]   Chen JQ, Delannoy M, Cooke C, Yager JD. Mitochondrial localization of ERalpha and ERbeta in human MCF7 cells. Am J Physiol Endocrinol Metab 2004; 286(6): E1011-22.
[http://dx.doi.org/10.1152/ajpendo.00508.2003] [PMID: 14736707]

[42]   Yang SH, Liu R, Perez EJ, *et al.* Mitochondrial localization of estrogen receptor beta. Proc Natl Acad Sci USA 2004; 101(12): 4130-5.
        [http://dx.doi.org/10.1073/pnas.0306948101] [PMID: 15024130]

[43]   Milanesi L, Vasconsuelo A, de Boland AR, Boland R. Expression and subcellular distribution of native estrogen receptor beta in murine C2C12 cells and skeletal muscle tissue. Steroids 2009; 74(6): 489-97.
        [http://dx.doi.org/10.1016/j.steroids.2009.01.005] [PMID: 19428437]

[44]   Vasconsuelo A, Milanesi L, Boland R. 17Beta-estradiol abrogates apoptosis in murine skeletal muscle cells through estrogen receptors: role of the phosphatidylinositol 3-kinase/Akt pathway. J Endocrinol 2008; 196(2): 385-97.
        [http://dx.doi.org/10.1677/JOE-07-0250] [PMID: 18252962]

[45]   La Colla A, Vasconsuelo A, Boland R. Estradiol exerts antiapoptotic effects in skeletal myoblasts *via* mitochondrial PTP and MnSOD. J Endocrinol 2013; 216(3): 331-41.
        [http://dx.doi.org/10.1530/JOE-12-0486] [PMID: 23213199]

[46]   La Colla A, Boland R, Vasconsuelo A. 17b-Estradiol abrogates apoptosis inhibiting PKCd, JNK, and p66Shc activation in C2C12 cells. J Cell Biochem 2015; 116(7): 1454-65.
        [http://dx.doi.org/10.1002/jcb.25107] [PMID: 25649128]

[47]   Pronsato L, Boland R, Milanesi L. Testosterone exerts antiapoptotic effects against H_2O_2 in C2C12 skeletal muscle cells through the apoptotic intrinsic pathway. J Endocrinol 2012; 212(3): 371-81.
        [http://dx.doi.org/10.1530/JOE-11-0234] [PMID: 22219300]

[48]   Viña J, Borrás C, Gambini J, Sastre J, Pallardó FV. Why females live longer than males? Importance of the upregulation of longevity-associated genes by oestrogenic compounds. FEBS Lett 2005; 579(12): 2541-5.
        [http://dx.doi.org/10.1016/j.febslet.2005.03.090] [PMID: 15862287]

[49]   Zhang L, Wu S, Ruan Y, Hong L, Xing X, Lai W. Testosterone suppresses oxidative stress *via* androgen receptor-independent pathway in murine cardiomyocytes. Mol Med Rep 2011; 4(6): 1183-8.
        [PMID: 21785825]

[50]   Borrás C, Gambini J, López-Grueso R, Pallardó FV, Viña J. Direct antioxidant and protective effect of estradiol on isolated mitochondria. Biochim Biophys Acta 2010; 1802(1): 205-11.
        [http://dx.doi.org/10.1016/j.bbadis.2009.09.007] [PMID: 19751829]

[51]   Roubenoff R. Catabolism of aging: is it an inflammatory process? Curr Opin Clin Nutr Metab Care 2003; 6(3): 295-9.
        [http://dx.doi.org/10.1097/01.mco.0000068965.34812.62] [PMID: 12690262]

[52]   Carmeli E, Coleman R, Reznick AZ. The biochemistry of aging muscle. Exp Gerontol 2002; 37(4): 477-89.
        [http://dx.doi.org/10.1016/S0531-5565(01)00220-0] [PMID: 11830351]

[53]   Zembroń-Łacny A, Dziubek W, Rogowski Ł, Skorupka E, Dąbrowska G. Sarcopenia: monitoring, molecular mechanisms, and physical intervention. Physiol Res 2014; 63(6): 683-91.
        [PMID: 25157651]

[54]   Argilés JM, López-Soriano FJ, Busquets S. Apoptosis signalling is essential and precedes protein degradation in wasting skeletal muscle during catabolic conditions. Int J Biochem Cell Biol 2008; 40(9): 1674-8.
        [http://dx.doi.org/10.1016/j.biocel.2008.02.001] [PMID: 18329944]

[55]   Hasselgren PO, Wray C, Mammen J. Molecular regulation of muscle cachexia: it may be more than the proteasome. Biochem Biophys Res Commun 2002; 290(1): 1-10.
        [http://dx.doi.org/10.1006/bbrc.2001.5849] [PMID: 11779124]

[56]   Purintrapiban J, Wang MC, Forsberg NE. Degradation of sarcomeric and cytoskeletal proteins in cultured skeletal muscle cells. Comp Biochem Physiol B Biochem Mol Biol 2003; 136(3): 393-401.

[http://dx.doi.org/10.1016/S1096-4959(03)00201-X] [PMID: 14602148]

[57]   Dirks A, Leeuwenburgh C. Apoptosis in skeletal muscle with aging. Am J Physiol Regul Integr Comp Physiol 2002; 282(2): R519-27.
[http://dx.doi.org/10.1152/ajpregu.00458.2001] [PMID: 11792662]

[58]   Dupont-Versteegden EE. Apoptosis in muscle atrophy: relevance to sarcopenia. Exp Gerontol 2005; 40(6): 473-81.
[http://dx.doi.org/10.1016/j.exger.2005.04.003] [PMID: 15935591]

[59]   Dirks AJ, Leeuwenburgh C. Aging and lifelong calorie restriction result in adaptations of skeletal muscle apoptosis repressor, apoptosis-inducing factor, X-linked inhibitor of apoptosis, caspase-3, and caspase-12. Free Radic Biol Med 2004; 36(1): 27-39.
[http://dx.doi.org/10.1016/j.freeradbiomed.2003.10.003] [PMID: 14732288]

[60]   Marzetti E, Carter CS, Wohlgemuth SE, *et al.* Changes in IL-15 expression and death-receptor apoptotic signaling in rat gastrocnemius muscle with aging and life-long calorie restriction. Mech Ageing Dev 2009; 130(4): 272-80.
[http://dx.doi.org/10.1016/j.mad.2008.12.008] [PMID: 19396981]

[61]   Jang YC, Lustgarten MS, Liu Y, *et al.* Increased superoxide *in vivo* accelerates age-associated muscle atrophy through mitochondrial dysfunction and neuromuscular junction degeneration. FASEB J 2010; 24(5): 1376-90.

[62]   Brioche T, Pagano AF, Py G, Chopard A. Muscle wasting and aging: Experimental models, fatty infiltrations, and prevention. Mol Aspects Med 2016; 50: 56-87.
[http://dx.doi.org/10.1016/j.mam.2016.04.006] [PMID: 27106402]

[63]   Siu PM, Pistilli EE, Alway SE. Apoptotic responses to hindlimb suspension in gastrocnemius muscles from young adult and aged rats. Am J Physiol Regul Integr Comp Physiol 2005; 289(4): R1015-26.
[http://dx.doi.org/10.1152/ajpregu.00198.2005] [PMID: 15919734]

[64]   Baker DJ, Hepple RT. Elevated caspase and AIF gene expression correlate with progression of sarcopenia during aging in male F344BN rats. Exp Gerontol 2006; 41(11): 1149-56.
[http://dx.doi.org/10.1016/j.exger.2006.08.007] [PMID: 17029665]

[65]   Otrocka-Domagała I. Sensitivity of skeletal muscle to pro-apoptotic factors. Pol J Vet Sci 2011; 14(4): 683-94.
[http://dx.doi.org/10.2478/v10181-011-0104-x] [PMID: 22439346]

[66]   Adhihetty PJ, O'Leary MF, Hood DA. Mitochondria in skeletal muscle: adaptable rheostats of apoptotic susceptibility. Exerc Sport Sci Rev 2008; 36(3): 116-21.
[http://dx.doi.org/10.1097/JES.0b013e31817be7b7] [PMID: 18580291]

[67]   Campion DR. The muscle satellite cell: a review. Int Rev Cytol 1984; 87: 225-51.
[http://dx.doi.org/10.1016/S0074-7696(08)62444-4] [PMID: 6370890]

[68]   Grounds MD, White JD, Rosenthal N, Bogoyevitch MA. The role of stem cells in skeletal and cardiac muscle repair. J Histochem Cytochem 2002; 50(5): 589-610.
[http://dx.doi.org/10.1177/002215540205000501] [PMID: 11967271]

[69]   Mauro A. Satellite cell of skeletal muscle fibers. J Biophys Biochem Cytol 1961; 9: 493-5.
[http://dx.doi.org/10.1083/jcb.9.2.493] [PMID: 13768451]

[70]   Christov C, Chrétien F, Abou-Khalil R, *et al.* Muscle satellite cells and endothelial cells: close neighbors and privileged partners. Mol Biol Cell 2007; 18(4): 1397-409.
[http://dx.doi.org/10.1091/mbc.e06-08-0693] [PMID: 17287398]

[71]   Ryall JG, Schertzer JD, Lynch GS. Cellular and molecular mechanisms underlying age-related skeletal muscle wasting and weakness. Biogerontology 2008; 9(4): 213-28.
[http://dx.doi.org/10.1007/s10522-008-9131-0] [PMID: 18299960]

[72]   Chargé SB, Rudnicki MA. Cellular and molecular regulation of muscle regeneration. Physiol Rev

2004; 84(1): 209-38.
[http://dx.doi.org/10.1152/physrev.00019.2003] [PMID: 14715915]

[73] Shamim B, Hawley JA, Camera DM. Protein availability and satellite cell dynamics in skeletal muscle. Sports Med 2018; 48(6): 1329-43.
[http://dx.doi.org/10.1007/s40279-018-0883-7] [PMID: 29557519]

[74] Nnodim JO. Satellite cell numbers in senile rat levator ani muscle. Mech Ageing Dev 2000; 112(2): 99-111.
[http://dx.doi.org/10.1016/S0047-6374(99)00076-7] [PMID: 10690923]

[75] Kadi F, Charifi N, Denis C, Lexell J. Satellite cells and myonuclei in young and elderly women and men. Muscle Nerve 2004; 29(1): 120-7.
[http://dx.doi.org/10.1002/mus.10510] [PMID: 14694507]

[76] Chakkalakal JV, Jones KM, Basson MA, Brack AS. The aged niche disrupts muscle stem cell quiescence. Nature 2012; 490(7420): 355-60.
[http://dx.doi.org/10.1038/nature11438] [PMID: 23023126]

[77] Van der Meer SF, Jaspers RT, Jones DA, Degens H. Time-course of changes in the myonuclear domain during denervation in young-adult and old rat gastrocnemius muscle. Muscle Nerve 2011; 43: 212e22.
[http://dx.doi.org/10.1002/mus.21822]

[78] Brack AS, Conboy MJ, Roy S, *et al.* Increased Wnt signaling during aging alters muscle stem cell fate and increases fibrosis. Science 2007; 317: 807e10.
[http://dx.doi.org/10.1126/science.1144090]

[79] Fry CS, Lee JD, Mula J, *et al.* Inducible depletion of satellite cells in adult, sedentary mice impairs muscle regenerative capacity without affecting sarcopenia. Nat Med 2015; 21(1): 76-80.
[http://dx.doi.org/10.1038/nm.3710] [PMID: 25501907]

[80] Fukada SI. The roles of muscle stem cells in muscle injury, atrophy and hypertrophy. J Biochem 2018; 163(5): 353-8.
[http://dx.doi.org/10.1093/jb/mvy019] [PMID: 29394360]

[81] Zammit P, Beauchamp J. The skeletal muscle satellite cell: stem cell or son of stem cell? Differentiation 2001; 68(4-5): 193-204.
[http://dx.doi.org/10.1046/j.1432-0436.2001.680407.x] [PMID: 11776472]

[82] Moss FP, Leblond CP. Satellite cells as the source of nuclei in muscles of growing rats. Anat Rec 1971; 170(4): 421-35.
[http://dx.doi.org/10.1002/ar.1091700405] [PMID: 5118594]

[83] Schultz E. Satellite cell proliferative compartments in growing skeletal muscles. Dev Biol 1996; 175(1): 84-94.
[http://dx.doi.org/10.1006/dbio.1996.0097] [PMID: 8608871]

[84] Beauchamp JR, Heslop L, Yu DS, *et al.* Expression of CD34 and Myf5 defines the majority of quiescent adult skeletal muscle satellite cells. J Cell Biol 2000; 151(6): 1221-34.
[http://dx.doi.org/10.1083/jcb.151.6.1221] [PMID: 11121437]

[85] Carosio S, Berardinelli MG, Aucello M, Musarò A. Impact of ageing on muscle cell regeneration. Ageing Res Rev 2011; 10(1): 35-42.
[http://dx.doi.org/10.1016/j.arr.2009.08.001] [PMID: 19683075]

[86] Doumit ME, Cook DR, Merkel RA. Testosterone up-regulates androgen receptors and decreases differentiation of porcine myogenic satellite cells *in vitro*. Endocrinology 1996; 137(4): 1385-94.
[http://dx.doi.org/10.1210/endo.137.4.8625915] [PMID: 8625915]

[87] Sinha-Hikim I, Taylor WE, Gonzalez-Cadavid NF, Zheng W, Bhasin S. Androgen receptor in human skeletal muscle and cultured muscle satellite cells: up-regulation by androgen treatment. J Clin Endocrinol Metab 2004; 89(10): 5245-55.

[http://dx.doi.org/10.1210/jc.2004-0084] [PMID: 15472231]

[88]　Yaffe D, Saxel O. Serial passaging and differentiation of myogenic cells isolated from dystrophic mouse muscle. Nature 1977; 270(5639): 725-7.
[http://dx.doi.org/10.1038/270725a0] [PMID: 563524]

[89]　Blau HM, Chiu CP, Webster C. Cytoplasmic activation of human nuclear genes in stable heterocaryons. Cell 1983; 32(4): 1171-80.
[http://dx.doi.org/10.1016/0092-8674(83)90300-8] [PMID: 6839359]

[90]　Patz TM, Doraiswamy A, Narayan RJ, *et al.* Two dimensional differential adherence and alignment of C2C12 myoblasts. Mater Sci Eng 2005; 123: 242-7.
[http://dx.doi.org/10.1016/j.mseb.2005.08.088]

[91]　Yoshida N, Yoshida S, Koishi K, Masuda K, Nabeshima Y. Cell heterogeneity upon myogenic differentiation: down-regulation of MyoD and Myf-5 generates 'reserve cells'. J Cell Sci 1998; 111(Pt 6): 769-79.

[92]　Milanesi L, Russo de Boland A, Boland R. Expression and localization of estrogen receptor alpha in the C2C12 murine skeletal muscle cell line. J Cell Biochem 2008; 104(4): 1254-73.
[http://dx.doi.org/10.1002/jcb.21706] [PMID: 18348185]

[93]　Vasconsuelo A, Pronsato L, Ronda AC, Boland R, Milanesi L. Role of 17β-estradiol and testosterone in apoptosis. Steroids 2011; 76(12): 1223-31.
[http://dx.doi.org/10.1016/j.steroids.2011.08.001] [PMID: 21855557]

[94]　Pronsato L, Ronda AC, *et al.* Protective role of 17b-estradiol and testosterone in apoptosis of skeletal muscle. Actual Osteol 2010; 2: 45-8.

[95]　Pronsato L. Role of testosterone and its receptor in the apoptosis of murine skeletal muscle cells. Doctoral thesis 2014; 1-165.

[96]　Reebye V, Querol Cano L, Lavery DN, *et al.* Role of the HSP90-associated cochaperone p23 in enhancing activity of the androgen receptor and significance for prostate cancer. Mol Endocrinol 2012; 26(10): 1694-706.
[http://dx.doi.org/10.1210/me.2012-1056] [PMID: 22899854]

[97]　Zoubeidi A, Zardan A, Beraldi E, *et al.* Cooperative interactions between androgen receptor (AR) and heat-shock protein 27 facilitate AR transcriptional activity. Cancer Res 2007; 67(21): 10455-65.
[http://dx.doi.org/10.1158/0008-5472.CAN-07-2057] [PMID: 17974989]

[98]　Sinha-Hikim I, Braga M, Shen R, Sinha Hikim AP. Involvement of c-Jun NH2-terminal kinase and nitric oxide-mediated mitochondria-dependent intrinsic pathway signaling in cardiotoxin-induced muscle cell death: role of testosterone. Apoptosis 2007; 12(11): 1965-78.
[http://dx.doi.org/10.1007/s10495-007-0120-6] [PMID: 17786558]

[99]　Renault V, Thornell LE, Butler-Browne G, Mouly V. Human skeletal muscle satellite cells: aging, oxidative stress and the mitotic clock. Exp Gerontol 2002; 37(10-11): 1229-36.
[http://dx.doi.org/10.1016/S0531-5565(02)00129-8] [PMID: 12470836]

[100]　Kayo T, Allison DB, Weindruch R, Prolla TA. Influences of aging and caloric restriction on the transcriptional profile of skeletal muscle from rhesus monkeys. Proc Natl Acad Sci USA 2001; 98(9): 5093-8.
[http://dx.doi.org/10.1073/pnas.081061898] [PMID: 11309484]

[101]　Vasconsuelo A, Milanesi L, Boland R. Actions of 17β-estradiol and testosterone in the mitochondria and their implications in aging. Ageing Res Rev 2013; 12(4): 907-17.
[http://dx.doi.org/10.1016/j.arr.2013.09.001] [PMID: 24041489]

[102]　Lim JJ, Ngah WZ, Mouly V, Abdul Karim N. Reversal of myoblast aging by tocotrienol rich fraction posttreatment. Oxid Med Cell Longev 2013; 2013: 978101.
[http://dx.doi.org/10.1155/2013/978101] [PMID: 24349615]

CHAPTER 8

The Role of Phytoestrogens in Apoptosis: Chemical Structures and Actions on Specific Receptors

María Belén Faraoni[*] and **Florencia Antonella Musso**

Instituto de Quimica del Sur (INQUISUR), Universidad Nacional del Sur- CONICET, Bahia Blanca, Argentina

Abstract: Phytoestrogens are polyphenolic nonsteroidal plant compounds with have estrogen-like biological activity. According to their chemical structures, phytoestrogens might be organised into three central groups: flavonoids, lignans and stilbenes. Isoflavonoids, a subgroup of flavonoids, are the most studied ones for their biological activities, and they are present in many foods, such as soybeans. The most representative isoflavonoids are genistein and daidzein. Due to the fact that phytoestrogens are considerably similar in structure to estrogen17β-estradiol, they may display selective estrogen receptors (ERs) modulating activities; having a higher affinity for ERβ than for ERα. Several studies conducted in animals and humans have indicated that one of the main functions of phytoestrogens involves having a protective effect on certain conditions which are estrogen-dependent, such as symptoms related to menopause, and on estrogen-dependent diseases including prostate and breast cancer, osteoporosis and heart disease. However, phytoestrogens have also anti-estrogenic properties, which have raised concerns since they might cause adverse health effects. At the moment, the existing data are not enough to support a more sophisticated semiquantitative risk-benefit analysis. Hence, phytoestrogens are currently being studied for their role in human health.

Keywords: 17β-Estradiol, Apoptosis, Bone Tissue, Daidzein, Estrogen Receptors, Genistein, Hormonal Replacement Therapy, Hormone-Dependent Cancer, Isoflavones, Menopause, Osteoporosis, Phytoestrogens, Postmenopausal Women, Proliferation Cell, Protective Effects, Sarcopenia, Skeletal Muscle, Soybeans.

INTRODUCTION

Plants produce a considerable amount of compounds which are categorized into primary and secondary metabolites. The former play a crucial part in photosynthesis, respiration, and the flourishing of plants. The latter, which are not

[*] **Corresponding author Maria Belén Faraoni:** Instituto de Quimica del Sur (INQUISUR), Universidad Nacional del Sur- CONICET, Bahia Blanca, Argentina; Research Member of CIC. E-mail: bfaraoni@criba.edu.ar

Andrea Vasconsuelo (Ed.)

produced by humans in a natural way and have no nutritional value, can be decisive in the preservation or destruction of human health. They can be found in a restricted amount of plants and they have varied structures. They are also a likely resource of natural biologically active compounds [1]. One group of secondary metabolites is constituted by phytoestrogens. They can be found in nuts, oilseeds, soy, and other foods.

Having a similar structure to the one present in female sex hormones, especially with 17β-estradiol, phytoestrogens symbolize the group of organic products which bear a resemblance to the biological activity of estrogens. Investigations show that through the consumption of phytoestrogens and foods abundant in these compounds, there can be a protective effect on certain conditions which are estrogen-dependent, such as symptoms related to menopause, and on estrogen-dependent diseases including prostate and breast cancer, osteoporosis and heart disease [2].

Many researches have reported the advantages of phytoestrogen consumption and the benefits produced on people who have a diet rich in phytoestrogens. However, phytoestrogens have also anti-estrogenic properties, which have raised concerns since they might cause adverse health effects. At the moment, the existing data are not enough to support a more sophisticated semiquantitive risk-benefit analysis. Hence, phytoestrogens are currently being studied for their role in human health.

In this chapter, we will present the classification of phytoestrogens, the importance of their chemical structures, and how they are found in nature. Furthermore, we will evaluate the role of phytoestrogens in skeletal and bone tissues and their effects on apoptosis in skeletal muscle cells. With this aim, we searched and discussed recent literature on phytoestrogens in different databases, focusing on their chemical features and their capacity of apoptosis in muscle and bone tissue and on the experimental studies performed in animals and humans.

Phytoestrogens Classification and Chemical Structure

Phytoestrogens are polyphenolic nonsteroidal plant compounds which can act like estrogen and develop normally in most plants, fruits, and vegetables. As they have a phenolic ring, they can bind to estrogen receptors in humans. These secondary metabolites are prevalent in more than 300 different plant species, but most of them are contained in leguminous. On the basis of their chemical structures, phytoestrogens might be organized into three central groups: flavonoids, lignans and, stilbenes (Fig. **1**) [Review in 2].

Flavonoids are broadly present in plants. Over four thousand ones based on the C_6-C_3-C_6 skeleton have been characterized and grouped into several classes, depending on the carbon of the C ring on which the B ring is attached and on the degree of unsaturation and oxidation of the C ring. Flavonoids in which the B ring is linked in position 3 of the C ring are called isoflavonoids. Both isoflavones, which are the most studied group of phytoestrogens, and coumestans, which possess the most distinct estrogenic effect of all phytoestrogens, belong to this class of isoflavonoids. On the other hand, the flavonoids with the basic skeleton (B ring linked in position 2) can be further subdivided into numerous subclasses according to the structural features of the C ring, such as flavones, flavanones, flavonols, flavanonols, chalcones and anthocyanidins.

The second group, lignans, were identified within the plants in which they assist in the production of lignin, used to build the plant cell wall. Lignans are biologically active phenolic compounds of plant origin with a poor estrogenic activity. They have two phenylpropane units (C_6-C_3) connected *via* two specific carbons (C-2–C-2') in their structure.

Stilbenes are plant phenolic compounds with a C_6-C_2-C_6 skeleton. The main representative one is resveratrol. Although there are two isomers of resveratrol, *cis*- and *trans*-resveratrol, only the *trans*- form has been reported to be estrogenic. Stilbene monomers, dimers and polymers occur widely in liverworts and higher plants [Review in 3, 4].

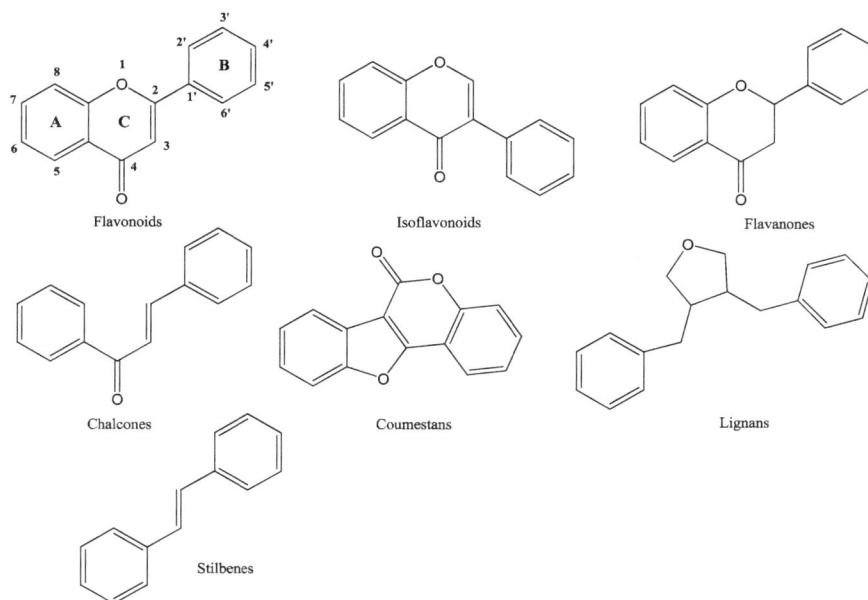

Fig. (1). Chemical structures of the main groups of phytoestrogens.

Phytoestrogens. Source and Biotransformation

Isoflavonoids

Out of all the groups of phytoestrogens, isoflavones are the ones which have been studied the most and appear almost solely in the Leguminosae family. A copious amount of isoflavones has been recognised from plants, with daidzein (**5**) and genistein (**6**) as the main ones. They manifest in plants as the inactive glycosides daidzin (**1**) and genistin (**2**) and their respectively 4'-methyl ether derivatives, formononetin (**3**) and biochanin A (**4**). Regardless of the great stability of the β-glycosides daidzin and genistin while being processed, these precursors might be metabolized in the digestive tract by the enzymes of the normal microflora to their corresponding aglycones, daidzein (**5**), and genistein (**6**). In the case of the gastro-intestinal microflora, it can further metabolize daidzein to the potent estrogen equol (**7**), but this biotransformation has high inter-individual variability. One more metabolite resulted from daidzein is *O*-demethylangolensin (O-DMA) (**8**). Metabolism of genistein by the microflora provides the end-products 2-(4-hydroxyphenyl)propanoic acid (**9**) and 1,3,5-trihydroxybenzene (**10**) (Fig. **2**). Finally, the metabolites of the isoflavones are excreted in urine [Review in 2].

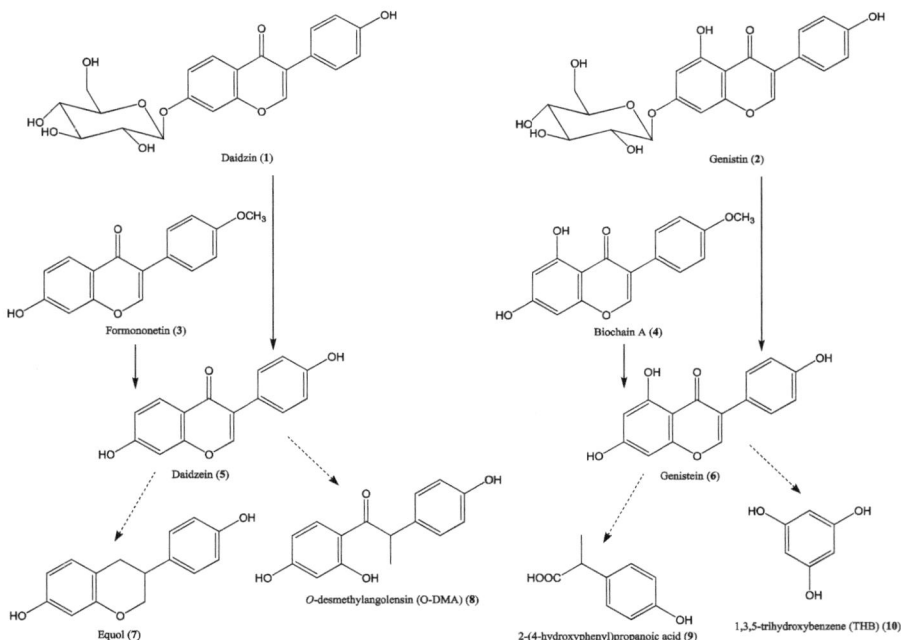

Fig. (2). Formation and biotransformation of the isoflavones daidzein (**5**) and genistein (**6**) (Extracted from [2]).

Isoflavonoids also include coumestans, with coumestrol (**11**) and 4'-*O*-methylcoumestrol (**12**) as the major members (Fig. **3**). When compared to isoflavones, coumestrol -known as a clover compound since 1957- has a 30 to 100 higher estrogenic activity. In addition, it can be located in spinach, brussels, sprout, and legumes like soybeans. Little is known about their metabolism in humans [Review in 2].

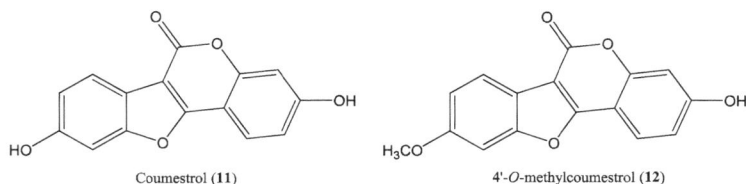

Fig. (**3**). Chemical structures of coumestrol (**11**) and 4'-*O*-methylcoumestrol (**12**).

Lignanes

The plant lignans may appear as aglycones and glycosides. They appear in oilseeds, such as flaxseed, but they also exist in whole cereals, grains, vegetables and fruits. Matairesinol (**13**) and secoisolariciresinol (**14**) have been recognized as two main plant precursors of mammalian lignans. Once they are ingested by intestinal bacterial flora, they are transformed into the biologically active metabolites enterolactone (**15**) and enterodiol (**16**), respectively (Fig. **4**). Both the parent compounds and the metabolites can be measured in several body fluids, such as urine, feces, and plasma [Review in 2].

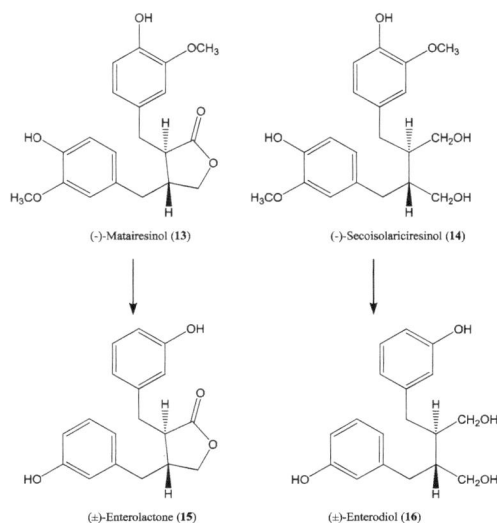

Fig. (**4**). Biotransformation of the lignans matairesinol (**13**) and secoisolariciresinol (**14**) (Extracted from [2]).

Stilbenes

Stilbens are found in grape-containing products, particularly red wine, in peanuts and peanuts products. The main representative one is resveratrol, being the *trans*-isomer (**17**) the most active. Not much is known about the biotransformation of stilbenes in humans.

The *trans*-resveratrol crosses the small intestine as such, but in the cell, this polyphenol undergoes rapid and extensive modifications by enzymes that produce glucuronides (**18** and **19**), and sulfates (**20** and **21**) conjugates (Fig. **5**). The metabolites leave the enterocyte to enter the systemic circulation. This unification to sulfates and glucoronides increments the aqueous solubility of the resveratrol, and decreases flux throughout membranes non-polar molecules from making contact with significant macromolecules and allows for excretion by the kidneys through urine. Consequently, the metabolism of resveratrol eventually leads to small quantities of free *trans*-resveratrol in the plasma to be sent to other tissues [5, 6].

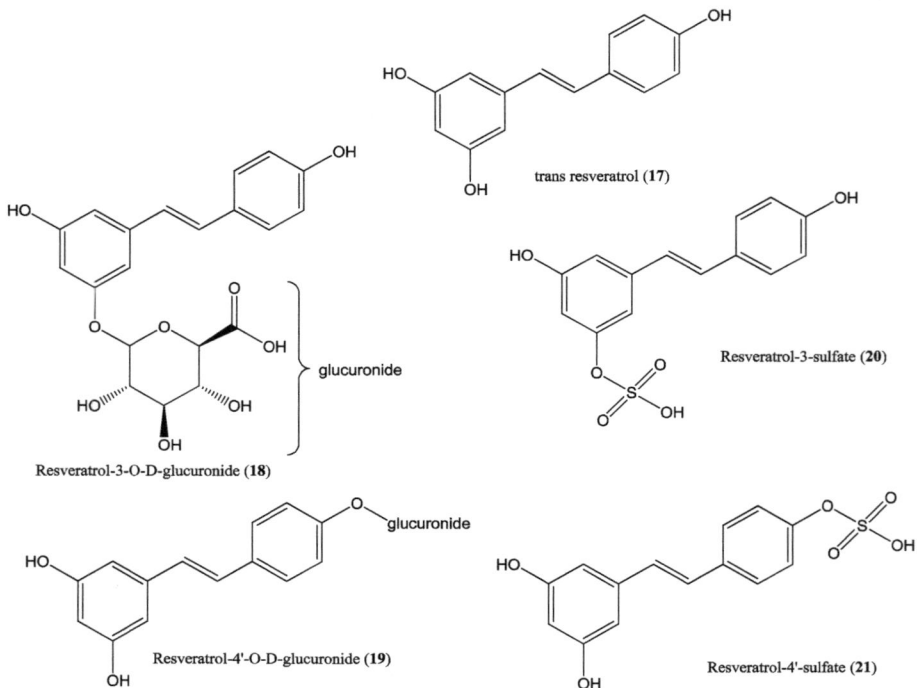

Fig. (5). Biotransformation of *trans*-resveratrol (**17**).

Phytoestrogens and Estrogen Receptors

As regards certain common features of all phytoestrogens, it can be said that their chemical structure is quite similar to that of 17β-estradiol (Fig. **6**) and that, in the body, they can provoke either estrogenic or anti-estrogenic response, depending on the concentration, the concentration of endogenous estrogen, and the distinctive features of an individual, mainly gender and hormonal status.

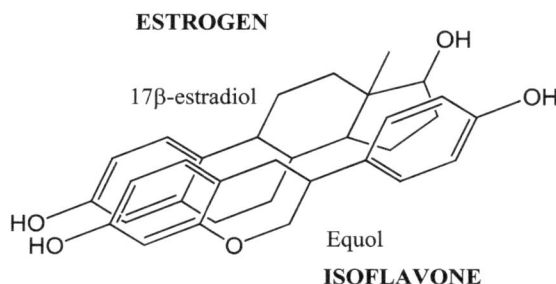

Fig. (6). Chemical structure similarity of estrogens and isoflavones.

Phytoestrogens exhibit these functions when binding to estrogen receptors (ERs) in tissues. The well-known "classical" estrogen receptors ERα and the newfound ERβ probably have different roles in gene regulation. How these receptors are present in several tissues varies. Thus, in rats the existence of ERα was observed in the uterus, testes, ovaries, kidneys, epididymis and thyroid gland; ERβ was mostly present in the prostate, lungs, bladder, and brain. In humans, ERα was seen in the testes, kidneys, and adrenal glands; both ERα and ERβ were detected in blood vessels, mammary glands, uterus, and ovaries, whereas only ERβ was found in the brain, lungs, thyroid gland, prostate, bladder, and bones.

It should be noted that the affinity which different estrogenic compounds have regarding the two types of estrogen receptors varies; particularly, phytoestrogens have a considerably higher affinity for ERβ than for ERα. Investigations on the occurrence and role of estrogen receptors in muscle tissue are quite new. Wiik *et al.* reported the existence of both kinds of receptors in the muscles of humans of both sexes and different ages. Their research showed the presence of ERs in the nuclei of muscle fibers and in the capillaries surrounding the fibers. This means that the muscle tissue is a target tissue for the estrogen action, as well as for those compounds that imitate estrogen, along with isoflavones. In humans, isoflavones are believed to be the most essential biologically active forms of phytoestrogens. Among them, daidzein (**5**) and genistein (**6**) are now being studied intensively for their effects on muscles (Fig. **2**) [Review in 7].

Such particular inclination for ERs allows isoflavones to exhibit estrogenic and antiestrogenic effects, determined by tissue type and endogenous estrogen content. For example, the antiestrogenic effect that genistein has on several cell lines is with an acceptable concentration of endogenous estradiol (1 nmol/dm^3). Moreover, genistein has a secondary estrogenic effect at lower concentrations of estradiol (0.01 nmol/dm^3), being common in postmenopausal women [Review in 4].

Interaction of Phytoestrogens with Estrogen Receptors

Estrogen receptors are mainly placed in the nuclear membrane. What results in the activation of what is known as estrogen response elements located on the inside of the nuclear membrane, is the interplay between phytoestrogens and ERs. Hence, the genomic mechanism, in certain transcription processes, are affected. Isoflavones act as agonists of ERs, but their activity is lower than that of 17β-estradiol (Table **1**). Their affinity is approximately 100-500 times lower than that of 17β-estradiol. Particularly, the affinity of genistein for ERβ is about 20-30 times higher than for ERα and is comparable to the affinity of 17β-estradiol. At satisfactorily high levels (over about 100 nmol/l for genistein), the effect of isoflavones could come close to the effect of endogenous 17β-estradiol at its physiological level. Considering that isoflavones and estradiol compete for their attachment to ERs, the effect of isoflavones is also dependent on the levels of endogenous estradiol. When dealing with high levels of endogenous estrogens, such as women in the follicular phase during the menstrual cycle, isoflavones might block full estrogen activity by filling up a part of the ERs. In contrast, during a condition with low indicators of endogenous estrogens, for instance, women in menopause or after ovariectomy, the estrogenic activity of isoflavones may happen. In this context, isoflavones are being highly used as a substitute or complement of hormonal replacement therapy (HRT) in postmenopausal women, particularly in those cases in which administration has been for a long time. The activation of membrane ERs leads to a cascade of intracellular mechanisms, which might comprise the control of the activity of G-proteins, adenylate cyclase, phospholipase or protein kinase. These cascades bring about quick effects on cell metabolism, along with changes in membrane permeability, ion concentration, production of nitric oxide (NO), *etc*. These processes are considered to play a significant part in tissues which are not regarded as traditional targets of the estradiol action [Review in 8].

Table 1. Relative estrogen activity in comparison with 17β-estradiol [4].

Compound	Relative activity of estrogen
17β-estradiol	100

(Table 1) cont.....

Compound	Relative activity of estrogen
coumestrol	0.202
genistein	0.084
equol	0.061
daidzein	0.013
formononetin	0.0006

Early studies of isoflavones focused on their estrogenic activity discovered that their actions are not entirely estrogenic and that their nonhormonal activities could be significant in the prevention of cancer. Some of the processes by which they might restrain cancer cells are: (1) inhibition of DNA topoisomerase, (2) suppression of angiogenesis, (3) induction of differentiation in cancer cell lines, and (4) induction of apoptosis. Many studies performed on *in vitro* cell culture and on *in vivo* animal experiments have exhibited that phytoestrogens can prevent tumors. In an extended review on the possibility of phytoestrogens minimizing the growth of tumors, Fournier *et al.* indicated that seventeen animal studies shown that adding soy products helped reduce tumor incidence or multiplicity in tumor models of the breast, prostate, liver, esophagus, and lung. This indicates that certain soy constituents are protective against cancer. Eleven of the reviewed studies examined isoflavone treatment (six with genistein, three with genistein/daidzein, one with biochanin-A, and another one with an unspecified isoflavone). Ten out of the eleven isoflavone studies reported a positive protective effect against cancer, whereas one resulted in enhanced tumor multiplicity.

Several studies main focus have been on the isoflavone genistein, which appears to be the main anticancer soy component. It contains antioxidant properties which can also contribute to its anticarcinogenic effects. It can inhibit hydrogen peroxide–induced tumor promoter activity *in vitro* and *in vivo*, and it has as well been shown to inhibit tyrosine kinase. The fact that it acts as an anticancer agent possibly arises from its suppression of enzymes that foster cell growth [Review in 9]. Currently, its effects on tumor cells are being investigated as a primary objective.

Effect of Phytoestrogens in Menopause

As is known, menopause consists of primordial transformations in the hormonal state and that these changes have a crucial impact on the density of bone mass and the distribution of body fat. Furthermore, a great amount of proof contributes to the hypothesis that the decrease in estrogen levels accompanied by menopause can lead to postmenopausal women losing muscle mass. The term commonly used to describe this muscle mass loss is known as sarcopenia.

It has been theorized that transitioning into menopause and the following decline in estrogen may contribute to muscle mass loss. For example, van Geel *et al.* expressed a positive relation concerning lean body mass and estrogen levels. Likewise, Iannuzzi-Sucich *et al.* found that muscle mass is remarkably related to plasma estrone and estradiol levels in women.

Since it's not yet understood the processes through which a decrease in estrogen levels might have a negative effect on muscle mass, it has been proposed that the said reduction may be connected to a rise in pro-inflammatory cytokines, namely tumor necrosis factor alpha (TNF-α) or interleukine-6 (IL-6), which might be involved in the manifestation of sarcopenia. In addition, estrogens might have an immediate effect on muscle mass given that it has been proved that skeletal muscle has estrogen beta-receptors on the cell membrane, in the cytoplasm, and on the nuclear membrane. Hence, a direct probable mechanistic link might be found between low estrogen levels and a decrease in protein synthesis. Other factors that lead to the progress of sarcopenia are shown in Fig. (**7**) [10].

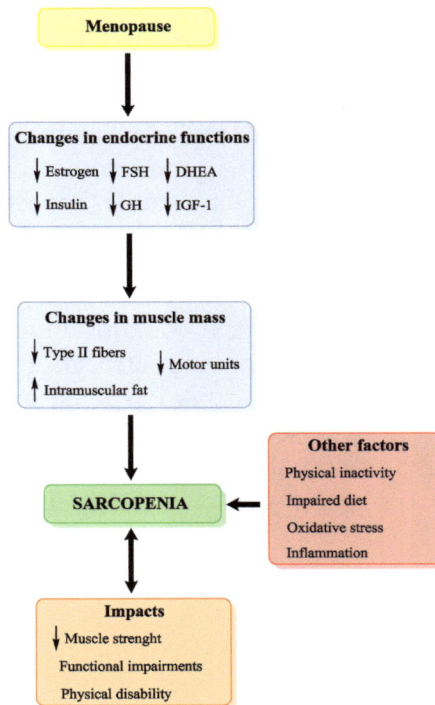

Fig. (7). Menopause-related changes on muscle mass and its impact on functional status (Extracted from [10]). FSH: follicle-stimulating hormone; DHEA: dehydroepiandrosterone; GH: growth hormone; IGF-1: insulin-like growth factor-1.

An additional potential method to countervail sarcopenia could be the addition of phytoestrogens. Isoflavone supplements are located in products containing soy; they apply a lipid-lowering effect, favor vasodilatation along with arterial compliance, and bring about the supervision of fasting glucose and insulin levels. Aubertin-Leheudre *et al.* studied the effect of a 70 mg/day soy isoflavone supplementation for 24 weeks on muscle mass in obese-sarcopenic postmenopausal women and realized that isoflavone supplementation was connected with a considerable rise in appendicular fat-free mass (+0.5 kg), even though this growth was not sufficient to reverse sarcopenia. As well, Moeller *et al.* randomized postmenopausal women to receive either isoflavone-rich soy protein (40 g) or isoflavone-poor soy protein or lacking eleven protein (control) for 24 weeks. It was revealed that changes in total lean body mass were not dissimilar among groups; nonetheless, lean body mass at the hip increased to a higher degree in the isoflavone-rich group (+3.4%) than in the isoflavone-poor (+1%) or control (0%) groups. At last, Maesta *et al.* calculated the effect of soy protein (25 g) along with resistance training on body composition in postmenopausal women. This research proved that soy protein together with 16 weeks of resistance training three times each week did not cause greater increases in muscle mass in comparison with resistance training alone, proposing that soy protein had no impact on muscle mass.

Estrogen supplementation or HRT is viewed as a likely strategy to play a protective part in muscle mass and muscle strength. For instance, Sorensen *et al.* carried out a 12-week double-blind study where either estrogen or placebo was provided and detected a considerable rise in lean body mass in women treated with estrogen. Furthermore, in the Women's Health Initiative study, volunteers who were randomized to get HRT for three years lost 0.04 kg of lean body mass, which was considerably less than the 0.44 kg lost by women on placebo, demonstrating that HRT could minimize muscle mass loss. In spite of that, certain investigations have not succeeded in showing a positive effect of HRT on muscle mass. For instance, in a study run by Hansen *et al.* the increase in muscle mass by given women 20 mg doses of estrogen for sixty-four weeks, was not significant. On the other hand, the frequency of sarcopenia was examined in women who had been on HRT for no less than two years. It was stated that they had a 23% incidence of sarcopenia though women not on HRT had a 22% incidence, implying that HRT does not avert the development of sarcopenia. Still, the contrasting outcomes between studies could be explicated by certain factors including the dose of estrogen administrated, length of the study, levels of training, dietary regimen, and medications. Another reason for the discrepancies could be that because of the contrasting stages of postmenopause when HRT was used, being the most favorable effects of HRT being in the early postmenopause period [Review in 10].

Some mechanistic signs for the favorable impact of HRT on human skeletal muscle mass and function have been given by Dieli-Conwright and colleagues. They exhibited that the expression of the genes furthering muscle growth (myogenic regulatory factors) was intensified, whereas the expression of the more negative regulators of muscle growth, like myostatin or proteolytic factors, was diminished in skeletal muscle of postmenopausal women using HRT, opposed to postmenopausal women not using HRT. The reduction of myostatin and activin IIb receptor expression was also higher after intense uncommon exercise in women using HRT, which can be one of the molecular mechanisms causing the effects of estrogens on muscle mass and strength preservation. Moreover, ER transcriptional activity, determined through messenger RNA (mRNA) expression of ER coregulators, was remarkably greater following uncommon exercise in skeletal muscle of women using HRT *versus* postmenopausal women not using HRT [Review in 11].

Although some studies showed that the HRT did not produce a beneficial result, more investigations stated that the effect was the opposite, that is, that HRT produced a benefit in the muscle mass of postmenopausal women. As it was already mentioned, the responses observed in the several studied treatments depend on various factors such as administered doses, age of women, postmenopausal stage, endogenous estrogen levels, exercise, diet, *etc.* Therefore, we can say that phytoestrogens produce a benefit in the muscular deficiencies developed in postmenopausal women, but it should be supported by further research. In addition, the consumption of soy phytoestrogen is associated with beneficial health properties similar to the ones proportioned by endogenous estrogens. The increase of occurrence of various diseases, such as hormone-dependent tumors, menopause symptoms, sarcopenia, *etc.*, is observed more in western countries than in eastern countries, especially Japan and China. It has been announced several times that Japan has the lowest risk of hormone-dependent cancer. People who migrate from Asia to western countries and keep their conventional diet, do not increment their risk of these diseases. Substantial dietary differences exist between these populations, mainly as regards the ingestion of soy products, which is said to be the highest one in some Japanese populations, with levels in the diet up to 200 mg/day. All through Asia, the estimated consumption of legumes supplies 25-45 mg/day of the total isoflavones compared with western countries, where less than 5 mg/day is consumed [Review in 12, 13]. The estrogenic effects of phytoestrogens may be responsible for changing the recurrence and intensity of the symptoms in these different populations.

Three central types of phytoestrogens can be found in foods: isoflavones (the most powerful), coumestans and lignans. The first ones appear in legumes such as soybeans, chickpeas, clover, lentils, and beans (Table **2**). The secondary soy products (milk or flour) have a lower quantity of isoflavones than the primary products. The coumestans, which are hardly ever ingested, appear in sprouting plants. The lignans are found in flaxseed (in vast amounts), lentils, whole grains, beans, fruits, and vegetables [Review in 12]. The presence of phytoestrogens in foods allows concluding that dietary intervention as a method of preventing the mentioned diseases is a pleasing and economical health benefit.

Table 2. Isoflavone contents in foods (µg/g fresh weight) [13].

Type of Food	Genistein	Daidzein	Glycitein	Biochanin A	Formononetin
Soybeans	335 – 1201	452 – 1138	37 – 145	< 1	< 1
Soy milk [mg/mL]	52 – 168	26 – 126	2 – 16	n.d.	n.d.
Tofu	111 – 304	73 – 191	15 – 39	n.d.	n.d.
Miso	51 – 398	35 – 363	4 – 53	n.i.	n.i.
Soy oil	n.d. – 3	n.d. – 1	n.d.	n.i.	n.i.
Soy sauce [mg/mL]	n.d. – 3	n.d. – 9	n.d. – 5	n.d.	n.d.
Soy flour	876 – 1155	715 – 1496	306 – 593	< 1	< 1
Soy protein isolate	272 – 1106	77 – 689	54 – 264	n.i.	n.i.
Tempeh	316 – 320	193 – 273	22 – 32	n.i.	n.i.
Natto	215 – 425	160 – 342	37 – 130	n.i.	n.i.
Soy cheese	3 – 150	3 – 98	3 – 53	n.i.	n.i.
Soy noodles	37 – 58	9 – 36	39	n.i.	n.i.
Soy bean sprouts	20	24	n.i.	n.d.	2
Clover sprouts	< 1	< 1	n.i.	8	40
Beans[a)]	n.d. – 7	n.d. – 0.2	n.d.	n.d.	n.d.
Garbanzo beans	< 1	n.d.	n.i.	14	0.5
Peas[a)]	n.d. – 53	n.d. – 73	n.i.	n.d.	n.d. – 93
Fruit, vegetables, nuts	0 – 2[b)]		n.d.	n.d.	n.d.

a) dry seeds, different varieties
b) total value;
n.d., below the given limit of detection; n.i., no information.

Effects of Phytoestrogens on Muscle Tissue

Estrogens improve the integrity of skeletal muscle cell membrane and decrease the loss of energy. Additionally, while the estrogenic compound estrone stimulates cell proliferation, 17β-estradiol (E2) has no impact on proliferation. Soy isoflavones consist of a phenolic ring and exhibit a specific agonistic and antagonistic properties on ERα and ERβ. It has been stated that genistein, daidzein, and glycitein have efficiently inhibited the proliferation of smooth muscle cells [Review in 12]. Different studies from several authors who have studied the effect of phytoestrogens, mainly isoflavones, on muscle tissue in diverse animal species are detailed below.

Jiang *et al.* studied the impact of various doses of isoflavones (0, 10, 20, 40, 80 mg/kg) on male broilers nourished with the identical basal diet without soybean meal. As a result, it was proven that by adding 10 and 20 mg/kg isoflavones the standard daily weight gain and food consumption was increased. Also, including 40 mg/kg isoflavones helped increase both the water holding capacity and the pH value of the meat. According to these outcomes, the authors concluded that including isoflavones in the diet of male broilers had a beneficial effect on growth and meat quality. Rehfeldt *et al.* researched the impacts of including isoflavone daidzein into the diet of sows amidst late gestation on the properties of the muscle tissue in the offspring. Conclusions of their study did not show statistically considerable differences in litter size, piglet birth weight, and percentage of muscle depending on the daidzein addition to the sow diet. Adding different amounts of daidzein to the diet of pregnant sows from day eighty-five of gestation up to the end of pregnancy had no implication on the muscle morphological features in newborn piglets and slaughter pigs when fattening ended. This study did not have statistically major effects on the histological characteristics of *musculus semitendinosus* neither in newborn piglets nor in slaughter pigs at the end of fattening. As regards the histological qualities of *musculus semitendinosus*, there was no dissimilarity in the size of the muscle cross-section, the sum of muscle fibers, and the number of nuclei per fiber, but a difference was noted in the presence of different fiber types within the muscle [Review in 7]. In spite of that, during tests with cultures of muscle cells (*in vitro*), it was shown that isoflavones inhibit the growth and development of muscle cells. Jones *et al.* discovered that phytoestrogens, particularly genistein, at a concentration of 1 mmol/l inhibits proliferation of rat muscle cells *in vitro* but does not alter protein degradation. Almost identical results were reached by Ji *et al.*, who discovered that genistein strongly inhibits proliferation and fusion of myoblasts in rats, that the inhibition strength is dependent on the dose of genistein, and that the effective dose is already from 1 µmol/l. Authors did not observe adverse effects of genistein on protein degradation [Review in 12]. The direct effects of genistein and daidzein on the proliferation of muscle cell cultures originating from newborn piglets were studied by Mau *et al.* During this investigation, doses of 0.1, 1, 10 and 100 µM of isoflavones were used; they would be calculated in the serum after consuming the milk-based diet of soybeans for infants or plant foods that are generally used in the diet for pigs. The results of this investigation revealed that the effects of isoflavones depended on the dose of isoflavones used and that genistein had a stronger inhibitory effect compared to daidzein on the proliferation of muscle cells. Such effect of genistein had already been demonstrated with a dose of 1 µM. In contrast, daidzein, even at considerably higher doses of up to 10 µM, did not have a detrimental effect on the growth of pig muscle cells [Review in 7]. Considering these facts, Rehfeldt *et al.* concluded that it has to be yet investigated if different doses and types of isoflavone compounds are more effective in influencing porcine muscle growth and meat quality, and that the inhibitory effect of genistein is contingent on the period of development of cell cultures [Review in 11].

Pueraria mirifica (PM) Airy Shaw & Suvatabandhu (Leguminosae) known in Thai as "Kwao Krua Kao," is a Thai medical phytoestrogen-rich plant. It is a traditional medicine used by menopausal women for rejuvenation and estrogen replacement. The plant tubers contain isoflavones (daidzin, daidzein, genistin, genistein, and puerarin) that constitute the majority of the active chemical ingredients of the tubers. The treatment with 40 mg/kg estrogen protected against skeletal muscle atrophy and encouraged a build-up of muscle strength and endurance; however, high levels of serum estrogen were found. In contrast, the treatment with 50, 500 or 1000 mg/kg PM also prevented muscle atrophy, restored muscle strength and endurance. It showed no toxicity to the hematopoietic system and the liver function. In addition, the serum estrogen was not too high. Thus, the estrous cycle was almost identical to the natural system after treatment with PM. However, treatment with 1000 mg/kg PM induced only two stages of the estrous cycle. Therefore, a 50 or 500 mg/kg PM treatment may be a possible dose for treatment of estrogen-dependent sarcopenia in ovariectomized rats. Treatment with 50, 500 or 1000 mg/kg PM for 90 days showed no toxicity, maintained the female rat reproductive system in ovariectomized rats in a natural manner, prevented muscle atrophy, and restored muscle strength and endurance [Review in 14]. These results may help to assess the possibility of using PM to replace estrogen therapy to prevent sarcopenia during menopause. The data from this study have demonstrated that PM is a potential choice to be used as HRT for improving skeletal muscle atrophy and loss of strength and endurance, especially in the estrogen-dependent sarcopenia.

The preceding information shows that, in nature, there are plant species rich in phytoestrogens capable of being studied for their possible use as HRT, lacking side effects and being as effective as the drugs that currently exist. Furthermore, many studies have demonstrated the ability of phytoestrogens to strengthen the skeletal muscle and inhibit cell proliferation, depending on the administered dose.

Effects of Phytoestrogens on Bone Tissue

The therapy involving estrogen replacement has an enormous effect in reducing the rate of bone loss and can in addition replace lost bone. If Japanese women, who have a diet abundant in soy products is compared with those having a western diet, the epidemiological evidence suggests that osteoporosis is about one third less in the former [Review in 9].

Isoflavone phytoestrogens promote the osteoblastic bone formation, inhibit the osteoclastic bone resorption, and prevent the excessive bone mass loss in mice and rats whose ovaries have been extirpated. The suggested method includes the stimulation of the osteoblast proliferation and the protection of osteoblasts from oxidative stress, as well as apoptosis of osteoclast progenitors. Clinical studies of the effects of phytoestrogens and foods that have phytoestrogens in bones have provided contrasting conclusions. It has been shown that supplementation with flax seed has no effect on the biomarkers of bone metabolism, while isoflavones avoid bone loss in postmenopausal women [Review in 4].

Low serum levels of 17β-estradiol are connected with lower calcium availability and activation of bone resorption-accelerating cytokines (IL-1, IL-6, IL-11, and TNF), which lead to the control of bone resorption over bone synthesis and following bone decalcification. Thus, osteoporosis is a significant difficulty in postmenopausal women. Even when estrogens assigned for short periods of time enhance bone density. Moreover, providing isoflavones for long periods proved to have a great effect on bone metabolism. Six-month genistein administration to postmenopausal women resulted in a crucial increase in bone density and a simultaneous decrease in the concentration of biochemical markers of bone resorption. After 12-month genistein administration, the increment in bone density was consistent with the effects of estrogen HRT. Polkowski and Mazurek proposed that the beneficial impact of isoflavones on bone metabolism might be mediated by at least two mechanisms. The first one is the effect on osteoclasts *via* apoptosis is activated. The second one is the inhibition of tyrosine-kinase activity *via* modulation of membrane ERs with successive transformations in the activity of alkaline phosphatase. In agreement with this hypothesis, Blair *et al.* proclaimed that cell osteoclast cultures washed by a genistein concentrate presented decreased tyrosine-kinase activity with consequently reduced bone remodeling [Review in 8]. It has been shown through several small clinical trials in peri and postmenopausal women that there is a slight beneficial effect of isoflavones on bone. Dalais *et al.* found an increase of 5.2% in the bone mineral content of the whole body with three months of supplementation with 45 g of soy grits [Review in 15]. Potter *et al.* discovered a 2.2% increase in bone mineral density of the lumbar spine after six months of supplementation with soy protein isolate (40 g/day with 90 mg isoflavones), which was critical compared to a reduction of 0.6% with placebo [Review in 8]. Similarly, Alekel *et al*. noted that consumption of 40 g/day of soy protein isolate for six months, consisting of 80 mg of isoflavones, was important to help preserve bone mineral density of the lumbar spine (-0.2%) compared with the placebo (-1.3%). Clifton-Bligh *et al*. presented a dose-response on the bone mineral density of the proximal radius and the ulna that implemented isoflavones derived from clover for six months. The bone mineral density of the proximal radius and the ulna increased markedly by 4.1% with 57 mg/day and by 3.0% with 85.5 mg/day of isoflavones. The reaction with 28.5 mg/day of isoflavones was not significant [Review in 15]. Morabito *et al*. used genistein; in each of these investigations, almost 200 postmenopausal women were randomized to receive 44 mg/day of isoflavones, separated from red clover, or 99 mg/day of isoflavones contained in 25.6 g of soy protein isolate, or placebo for one year. Isoflavones significantly prevented bone loss by approximately 1% in the lumbar spine and by 1.3% in the intertrochanteric region of the hip compared to placebo [Review in 8].

Fewer studies have investigated the effects of flax lignans on bone mineral density, and for this reason, no evaluations can be made about their efficiency at this time. A study in growing female rats found some positive effects of flax lignans compared to a control diet, with an increase in femur strength at 50 days postnatal (adolescence) and there were no differences in bone mineral content. This proposes that flax lignans could have a beneficial effect on bone geometric properties and affect bone strength regardless of bone mineral content.

Moreover, phytoestrogens might work with exercise to affect bone. A cooperative effect on bone between exercise and estrogen may occur through their effects on the estrogen receptor. Exercise could generate bone formation *via* activation of the estrogen receptor, and this receptor is up-regulated by estrogen therapy. Estrogen receptor is a transcription factor for osteoblasts and controls gene expression when activated by estrogen or other ligands. It can appear in osteoclasts and osteocytes as well. Osteocytes are presumably associated with the identification of strains generated by exercise. The activation of estrogen receptor results in osteoblast proliferation and differentiation, promotion of osteocyte survival, and osteoclast apoptosis. Studies in animal and human osteoblast cell lines show that strain activates estrogen receptor, leading to osteoblast proliferation [Review in 15].

In this section of the chapter, it is very important to highlight that phytoestrogens are also capable of generating a protective effect on bone tissue. This occurs because of the relationship between muscle and bone, mentioned in previous chapters.

Effects of Phytoestrogens in Hormone-Dependent Cancer

As opposed to the population of western countries, it is known that the occurrence of some types of tumors is not as common in Asian countries. Factors related to the environment seem to have broadly contributed to the development of these tumors. Asian people who immigrate to western countries and modified their food choices (lower intake of soybeans and fiber sources, higher consumption of meat products) suffer more often from hormone-dependent cancer. This fact has been related to the capability of isoflavones to increment serum sex hormone binding globulin (SHBG) concentration, hence decreasing the bioavailability of sexual hormones in hormone-dependent tissues.

During a study, rat L6 skeletal muscle cells were used to evaluate soy phytoestrogen effects on cell proliferation, protein degradation, and protein synthesis. Genistein inhibited cellular proliferation in non–estrogen-induced cancer cells at concentrations of 1 μM and greater. This could happen because of the potential inhibitory effect of genistein on the tyrosine kinase receptor. Tyrosine kinases are a group of enzymes that have a central role in the formation and uncontrolled growth of cancer cells. At the starting point of the cell cycle, a two-subunit complex consists of cyclin B and Cdc2, the mitosis-promoting factor. This mitosis-promoting factor complex must be accurately phosphorylated and dephosphorylated on tyrosine and threonine sites for the cell cycle to progress. If these receptors are inhibited, mitosis does not occur. In the study, at a concentration of 0.7 μg/mL, genistein had a half-maximal effect on receptor tyrosine kinase inhibition [Review in 8].

By investigating the effect of estrogens *in vitro* in pig muscle cell cultures, Mau *et al*. detected hardly any effect on the proliferation of muscle cells at physiological concentrations, but this inhibitory effect was determined when the estrogen was administered in supraphysiological concentrations [Review in 7].

On the other hand, it has been stated that copious doses of genistein could stimulate cell proliferation in estrogen-dependent tumors. Contrastingly, Mertens *et al*. proclaimed that the supplying of isoflavones in the dose of 100 mg/day during the course of twelve months had no impact on cell proliferation in unaffected breast, unlike in women who had received treatment for breast cancer in the past. Likewise, Fabian *et al*. reported that isoflavones had no influence on the number of atypical cells present in the breast tissue. Thanks to these discoveries, great isoflavone intake is not recommended in patients with an identified diagnosis of estrogen-dependent breast cancer. In cell cultures, elevated doses of genistein and biochanin A inhibit the growth of prostate cancer cell lines [Review in 8].

Genistein is successful in inhibiting growth in cancer cells with and without stimulation by growth factors. It is so effective in estrogen-dependent breast cancer cell lines (MCF-7) and estrogen-independent (MDA-468), and is not altered by the existence of the multi-drug resistant gene product [Review in 8, 16, 17]. Genistein also acts *in vitro*, in an additive and/or synergistic manner together with other antitumor agents to inhibit the growth of cancer cells or to cause differentiation. In seventeen (65%) of the twenty six studies in animals, soybean or soy isoflavone consumption was found to decrease tumor progression, while no studies reported an increase in cancer cell growth [Review in 17].

Genistein and daidzein have been implicated in the prevention of cancer. Animal models of cancer and cell culture studies with human tumor-derived cell lines support this hypothesis.

The principal target of genistein in tumor cells has not yet been identified, although when in 1987 it was discovered that genistein is a distinct inhibitor of protein tyrosine kinases (PTKs) it resulted in its extensive use as a chemical probe to investigate the steps of signal transduction in both normal and transformed cell types. Along with its effects on PTKs that are part of signal transduction pathways, as it was already mentioned, genistein regulates events supervising the entry into and the way out of certain phases of the cell cycle, inhibits DNA topoisomerase II activity, modifies cell differentiation, and inhibits the production of reactive oxygen species [Review in 16].

It is necessary to be cautious when interpreting the available evidence. Several, but not all, of the tumor-inhibiting effects have been acquired with massive doses of phytoestrogens, much more than it could be gained by diet alone. Experimental conditions and isoflavone concentrations varied a lot. Cell lines were dissimilar and the presence or absence of estradiol differed as well. It is still not clearly known what doses and which kinds of phytoestrogens need to be used for tumor suppression, their continuation and periodicity, and the possible or real adverse effects or toxicities.

CONCLUDING REMARKS

Up to the present time, there are no intervention trials which use soy or its products in humans to prevent primary or secondary cancer. Even though these researches are necessary, they will be difficult to execute and control. Moreover, it is complicated to carry out retrospective epidemiological studies on soy intake. There are no recommendations that can be made in regards to the use of phytoestrogens in cancer prevention or treatment, although they appear to have hopeful effects *in vitro*. According to this reveals, no definite statement can be made about the protective effect of dietary phytoestrogens; regardless of this, there are animal studies and *in vitro* tests that propose that the soy constituents, especially the isoflavones, have antineoplastic activity. Therefore, after evaluating these studies performed on tumor cells and on cell proliferation, we observed that phytoestrogens can inhibit tumor growth and do not aggravate the pathology. Nevertheless, more studies are needed in the field of cancer and the use of phytoestrogens to confirm their possible use for the treatment of this disease. Given the data of potential adverse health effects, a definite conclusion cannot be made on the possible beneficial health effects of phytoestrogens.

CONSENT FOR PUBLICATION

Not applicable.

CONFLICT OF INTEREST

The authors confirm that this chapter contents have no conflict of interest.

ACKNOWLEDGEMENTS

National University of the South Argentina and National Research Council of Argentina (CONICET).

REFERENCES

[1]   Crozier A, Clifford MN, Ashihara H. Plant secondary metabolites: occurrence, structure and role in the human diet. John Wiley & Sons 2008.

[2]   Cos P, De Bruyne T, Apers S, Vanden Berghe D, Pieters L, Vlietinck AJ. Phytoestrogens: recent developments. Planta Med 2003; 69(7): 589-99.
[http://dx.doi.org/10.1055/s-2003-41122] [PMID: 12898412]

[3]   Cornwell T, Cohick W, Raskin I. Dietary phytoestrogens and health. Phytochemistry 2004; 65(8): 995-1016.
[http://dx.doi.org/10.1016/j.phytochem.2004.03.005] [PMID: 15110680]

[4]   Nikolić IL, Savić-Gajić IM, Tačić AD, Savić IM. Classification and biological activity of phytoestrogens: A review. Adv Technol 2017; 6: 96-106.
[http://dx.doi.org/10.5937/savteh1702096N]

[5]   Planas JM, Alfaras I, Colom H, Juan ME. The bioavailability and distribution of *trans*-resveratrol are constrained by ABC transporters. Arch Biochem Biophys 2012; 527(2): 67-73.
[http://dx.doi.org/10.1016/j.abb.2012.06.004] [PMID: 22750234]

[6]   Smoliga JM, Blanchard O. Enhancing the delivery of resveratrol in humans: if low bioavailability is the problem, what is the solution? Molecules 2014; 19(11): 17154-72.
[http://dx.doi.org/10.3390/molecules191117154] [PMID: 25347459]

[7]   Adamovic I, Vitorovi D, Petrovi M, Blagojevi M, Neši I. Influence of phytoestrogens on skeletal muscle structure. Proceedings of the International Symposium on Animal Science.

[8]   Pilšáková L, Riečanský I, Jagla F. The physiological actions of isoflavone phytoestrogens. Physiol Res 2010; 59(5): 651-64.
[PMID: 20406033]

[9]   Glazier MG, Bowman MA. A review of the evidence for the use of phytoestrogens as a replacement for traditional estrogen replacement therapy. Arch Intern Med 2001; 161(9): 1161-72.
[http://dx.doi.org/10.1001/archinte.161.9.1161] [PMID: 11343439]

[10]   Messier V, Rabasa-Lhoret R, Barbat-Artigas S, Elisha B, Karelis AD, Aubertin-Leheudre M. Menopause and sarcopenia: A potential role for sex hormones. Maturitas 2011; 68(4): 331-6.
[http://dx.doi.org/10.1016/j.maturitas.2011.01.014] [PMID: 21353405]

[11]   Velders M, Diel P. How sex hormones promote skeletal muscle regeneration. Sports Med 2013; 43(11): 1089-100.
[http://dx.doi.org/10.1007/s40279-013-0081-6] [PMID: 23888432]

[12]   Jones KL, Harty J, Roeder MJ, Winters TA, Banz WJ. *In vitro* effects of soy phytoestrogens on rat L6 skeletal muscle cells. J Med Food 2005; 8(3): 327-31.
[http://dx.doi.org/10.1089/jmf.2005.8.327] [PMID: 16176142]

[13]   Eisenbrand G. Isoflavones as phytoestrogens in food supplements and dietary foods for special medical purposes. Opinion of the Senate Commission on Food Safety (SKLM) of the German Research Foundation (DFG)-(shortened version) Mol Nutr Food Res 2007; 51: 1305-2.
[http://dx.doi.org/10.1002/mnfr.200700217]

[14]   Inthanuchit KS, Udomuksorn W, Kumarnsit E, Vongvatcharanon S, Vongvatcharanon U. Treatment with *Pueraria mirifica* extract prevented muscle atrophy and restored muscle strength in ovariectomized rats. Sains Malays 2017; 46: 1903-11.
[http://dx.doi.org/10.17576/jsm-2017-4610-29]

[15]   Chilibeck PD, Cornish SM. Effect of estrogenic compounds (estrogen or phytoestrogens) combined with exercise on bone and muscle mass in older individuals. Appl Physiol Nutr Metab 2008; 33(1): 200-12.
[http://dx.doi.org/10.1139/H07-140] [PMID: 18347673]

[16]   Barnes S, Grubbs C, Setchell KDR, Carlson J. Soybeans inhibit mammary tumors in models of breast cancer.Mutagens and Carcinogens in the Diet. New York: Wiley-Liss 1990; pp. 239-53.

[17]   Messina MJ, Persky V, Setchell KDR, Barnes S. Soy intake and cancer risk: a review of the *in vitro* and *in vivo* data. Nutr Cancer 1994; 21(2): 113-31.
[http://dx.doi.org/10.1080/01635589409514310] [PMID: 8058523]

CHAPTER 9

Conclusions

Andrea Vasconsuelo[*]

Instituto de Ciencias Biológicas y Biomédicas del Sur (INBIOSUR), Universidad Nacional del Sur- CONICET, Bahía Blanca, Argentina

Skeletal muscle atrophy and the loss of myofibers contribute to sarcopenia, a condition associated with normal aging. Certainly, the proapoptotic signaling is increased in muscles of old animals. Then the key mechanism involved in the skeletal muscle loss could be apoptosis. Since it is now well established, that skeletal muscle not only generate force and movement, apoptosis of this tissue could have a wide spectrum of effects on organism at different levels, as basal energy metabolism, in the storage for substrates, in the keep of core temperature, in glycemia, and in the use of oxygen and energy during movement. In addition, skeletal muscle acts as endocrine organ, thus could be regulated by own and no own hormones. Consequently, the progressive loss of skeletal muscle performance with aging is associated with functional impairments in daily life activities, the loss of independence, an increased risk of developing chronic metabolic diseases, and an overall decrease in the quality of life.

Clearly, discovering points of regulation of apoptosis process in skeletal muscle will lead to the development of therapies to deal with numerous pathologies. Although we are still far from understanding the events that occur first and trigger muscle apoptosis during aging, there is wide accord on the main role played by mitochondria. Interesting, sex steroid hormones may control mitochondrial functions by regulation of nuclear DNA encoded mitochondrial proteins, control of nuclear transcription factors affecting those nuclear DNA encoded mitochondrial proteins, regulation of mtDNA encoded proteins, activating signaling pathways through ARs or ERs, and by control mitochondrial cation fluxes regulating mitochondrial channels. These actions regulate major mitochondrial functions, such as apoptosis. In agreement, ARs and ERs was found in mitochondria, suggesting that both hormones could directly act on the organelle. The importance, the cytoskeleton acts as scaffolding of this signaling

---

[*] **Corresponding author Andrea A. Vasconsuelo:** Instituto de Ciencias Biológicas y Biomédicas del Sur (INBIOSUR), Universidad Nacional del Sur- CONICET, Bahía Blanca, Argentina.; E-mail: avascon@criba.edu.ar

system, supporting and interconnecting it; and this filamentous structure is susceptible to sex steroid regulation.

The co-existence of plasma membrane and mitochondrial ERs and ARs with the orthodox nuclear and cytoplasmic types of them, as well as the presence of additional proteins able to bind estradiol or testosterone and to trigger signaling responses, has changed our thinking about the hormone mechanism of action. Moreover, the ubiquitous localization of both receptors, suggests that this complex signaling network will be found in all body tissues. However, we focus on the action of sex steroids on skeletal muscle suggesting that the decrease of both hormones due different pathologies or during normal elderly is the responsible of sarcopenia and the specific mechanism responsible could be the deregulation of apoptosis.

Since the discovery of skeletal muscle satellite cells, a substantial body of scientific research has been dedicated to the properties and functions of these cells. The results of these studies have encouraged to state that the satellite cell has earned its place at the very centre of adult muscle physiology. The importance, the stem cells expressed both androgen and estrogen receptors, and they are responsive to sex steroids. The apoptosis induced by stress oxidative in satellite cells is abrogated by estradiol or testosterone.

Clearly, the protection of sex hormones against apoptosis on satellite cells decrease in the elderly, it could trigger sarcopenia. Pharmacologic inhibitors of critical second messenger systems have been developed to block apoptosis. It is imaginable that, in the future, therapeutic modalities inhibiting apoptotic processes, using the targets of sexual steroids to exert their beneficial effects, may have a role in minimizing the loss of satellite cells and then favoring the muscle regeneration/reparation. In this context, many investigations are being carried out with natural compounds and phytoestrogens, polyphenolic nonsteroidal plant compounds with estrogen-like biological activity, are good candidates to new therapies. Numerous studies in animals and humans have suggested that phytoestrogens can have a protective effect on estrogen-dependent conditions such as menopausal symptoms and on estrogen-dependent diseases. However, phytoestrogens have also anti-estrogenic properties, which have raised concerns since they might cause adverse health effects. Consequently, phytoestrogens are currently under active investigation for their role in human health.

CONSENT FOR PUBLICATION

Not applicable.

CONFLICT OF INTEREST

The authors confirm that this chapter contents have no conflict of interest.

ACKNOWLEDGEMENTS

National University of the South Argentina and National Research Council of Argentina (CONICET).

SUBJECT INDEX

A

Actin- binding proteins (ABPs) 122, 123
Actin cytoskeleton 120, 121, 122
Actin filaments 82, 120, 121, 122
Actin-interacting proteins function 36
Actin molecules 82
Actions 5, 7 8, 17, 82, 135, 136
  androgenic 17
  antiapoptotic 135, 136
  non-genomic 5, 7, 8
  non-genomic androgen 8
  physiological skeletal muscle 82
Activation function 2, 3
Activation function domain 64
Activation/phosphorylation 37, 96
Activation region 3
Activity 19, 26, 37, 42, 43, 83 123
  adipose tissue lipoprotein lipase 42
  biological 144, 145, 166
  low androgen/AR signaling 43
  myogenic 26
  physical 19, 37, 83
  severing 123
Adipose tissue 36, 42, 43
  white 42, 43
Adipose tissue mass 42
Adult skeletal muscle 133
Aglycones 147, 148
Amino acids 3, 80, 83, 92
Amygdala 17, 18, 19
Anabolic effects 12, 13, 26, 34, 38
Androgen action 1, 35
Androgen administration 25, 26, 30
Androgen/AR signaling involvement 39
Androgen binding entities 8
Androgen binding sites 9, 73, 75
Androgen concentrations 16, 38, 39, 40
  free 40
  high tubular 16
Androgen dependence 21
Androgen dependency 15
Androgen-dependent control 22
Androgen-dependent functions 16

Androgen depletion 123
Androgen deprivation 14, 31
Androgen deprivation therapy (ADT) 14, 31, 37, 43
Androgen effects 22, 23, 24, 34
Androgenetic alopecia 39, 40
Androgenic alopecia 39
Androgenic control 12, 24
Androgenic responsiveness 24
Androgenic steroids 26
Androgenic stimulation 16, 24
Androgen-induced hypervolemia 24
Androgen insensitivity 15
Androgen insensitivity mutations 15,18,32,39,41
Androgen levels 32
  high 39
  increased serum 41
  neonatal 18
Androgen occupancy 17
Androgen physiology 34
Androgen receptor action 14
Androgen receptor correlates 31
Androgen receptor overexpression 13
Androgen receptor signaling 15
Androgen receptor support 9
Androgen receptor translocation 9
Androgen replacement 15, 98
Androgen response 98
Androgen response elements (AREs) 3, 5
Androgens 1, 9, 14, 16, 21, 31, 41
  adrenal 39
  classical 9
  first-line 14
  low levels of 31
  neonatal 21
  peripheral 16
  physiological 1
  potent 41
Androgens acts 112
Androgens control apoptosis 112
Androgen sensitivity 39, 40
Androgen signaling 32
Androgens modulate 5, 41
Androgens orchestrate 12

www.ingramcontent.com/pod-product-compliance
Lightning Source LLC
Chambersburg PA
CBHW041704210326
41598CB00007B/521